小廚房滋味美食

Party recipes for the small kitchen

目錄
Contents

序言 / 4

沙律
Salad

泰式鮮果沙律 / 5
Fruit Salad in Thai Style

大蝦牛油果喀嗲 / 8
Prawn and Avocado Cocktail

柚子沙律 / 11
Pomelo Salad

三文魚南瓜芝士暖沙律 / 14
Warm Salad with Salmon, Pumpkin and Cheese

牛蒡沙律 / 17
Burdock Salad

吞拿魚沙律船 / 20
Cucumbers Stuffed with Tuna Salad

小食
Finger Food

烤焗一字排 / 23
Grilled Spareribs in Barbecue Sauce

香煎牛柳粒 / 26
Fried Beef Tenderloin Cubes

巴東牛肉 / 28
Beef Rendang

越式咖喱雞伴蒜茸包 / 31
Vietnamese Style Chicken Curry with Garlic Bread

西汁燴雞翼 / 34
Braised Chicken Wings in Western Sauce

蝦堡 / 36
Shrimp Burgers

泰式香辣雞翼 / 39
Thai Spicy Chicken Wings

番茄魔鬼蛋 / 42
Deviled Eggs with Tomatoes

番茄醃肉撻 / 44
Tomato Tart with Bacon

金粟菜肉煎餃 / 48
Sautéed Corn, Vegetable and Pork Dumplings

簡易麵包薄餅 / 51
Simple Bread Pizzas

墨西哥芝士紅腰豆薄餅 / 54
Kidney Bean and Cheese Burritos

芝蛋漢堡 / 56
Cheese and Egg Burgers

鷹嘴豆醬配比得包 / 58
Garbanzo Chickpea Paste with Pitta Bread

芝士酥條 / 60
Puff Pastry Cheese Pockets

香葱芝士球 / 63
Spring Onion Cheese Balls

迷你熱狗多士 / 66
Mini Hotdog Toasts

粉麵
Pasta

卡邦尼意粉 / 68
Carbonara Spaghetti

鰻魚蕎麥麵卷 / 70
Eel and Soba Sushi Rolls

香辣銀魚長通粉 / 72
Penne with Spicy Anchovy Fillet

甜品
Dessert

椰汁糕 / 75
Coconut Pudding

黑芝麻奶凍 / 78
Black Sesame Panna Cotta

藍莓芝士凍餅 / 80
No-bake Blueberry Cheesecakes

洛神花果凍 / 84
Hibiscus Tea Jelly

黑芝麻糖番薯 / 86
Candied Sweet Potato with Black Sesames

甜薯銅鑼燒 / 88
Sweet Potato Layered Dorayaki

陳皮紅豆沙 / 92
Red Bean Soup with Dried Tangerine Peel

提子麵包布甸 / 94
Raisin Bread Pudding

楓漿雪芳蛋糕 / 97
Maple Chiffon Cake

飲品
Drink

番茄西瓜汁 / 100
Tomato Watermelon Juice

檸檬薄荷茶 / 102
Mint Lemon Tea

百香果蜜汁 / 104
Honey Passionfruit Drink

三莓汁 / 106
Triple Berry Blend

烏豆糙米茶 / 108
Brown Rice Tea with Black Beans

鮮橙香草茶 / 110
Orange Herbal Tea

序言

Preface

小廚房也能煮出美食

在私人空間中，就算廚房是小小的，也能享受烹飪的樂趣。

其實烹調美食，只需要用基本的廚具就可以，書內食譜大多採用平底鑊、煲、小焗爐等，就能煮出一系列美食如沙律、小食、意粉、甜品和飲品。

聚會的目的，是享受閨蜜和好友的陪伴，互相傾訴大家的喜怒哀樂，毫無束縛。美食是聚會不可缺少的綠葉，書內挑選了一些容易處理、做法簡單的美食，還有助你零失敗的烹飪技巧，就算是入廚新手也可以成為聚會中的主廚。

The kitchen is small, but the food tastes big.

In the privacy of your own home, even though the kitchen is cramped, you still get to savour the joy of cooking.

Fancy cookware may save you time or effort. But you only need the bare kitchen essentials for most of the tasty dishes you wish to make. In this book, most recipes call for a pan, a pot or a small oven. With these tools, you can make a multi-course dinner with salad, snacks, pasta, desserts and even drinks.

Friends meet to enjoy the company of each other, and to update each other on their lives. You should be able to laugh as loud as you want, or cry as hard as you feel like. Nothing should hold you back. That's the beauty of a night-in. And food is indispensable to any party. The recipes compiled here are all easy and simple dishes that taste awesome. They come with fail-proof tips that help you nail them the first time around. Even kitchen newbies can take up the daunting task as the head chef in these parties.

泰式鮮果沙律

Fruit Salad in Thai Style

材料（4 人份量）

菠蘿 1/2 個
葡萄 100 克
火龍果 1 個
奇異果 2 個
蜜瓜 1/4 個

沙律汁

香茅 2 枝
糖 120 克
水 1/2 杯
青檸 1 個（磨皮茸、榨汁）

做法

1. 香茅切碎，與水、青檸皮茸、糖先用大火煲滾，再用慢火煮約 10 分鐘，待涼後加入青檸汁 1 湯匙，拌勻成沙律汁。
2. 菠蘿、火龍果、奇異果、蜜瓜去皮切粒；葡萄洗淨，切半去籽。
3. 將鮮果放於盆內，加入香茅糖水，拌勻放入雪櫃雪約數小時便可享用。

泰式鮮果沙律

Ingredients (Serves 4)

1/2 pineapple
100 g grapes
1 dragonfruit
2 kiwifruits
1/4 Hami melon

Salad dressing

2 stalks lemongrass
120 g sugar
1/2 cup water
1 lime (zest and juice)

Method

1. Finely chop lemongrass. Bring to the boil together with water, grated lime zest and sugar over high heat. Reduce to low heat and cook for 10 minutes. Leave to cool. Pour in 1 tbsp of lime juice. Mix well to make salad sauce.
2. Peel and dice pineapple, dragonfruit, kiwifruits and Hami melon. Wash grapes. Cut in half and seed.
3. Put the fruits in a big bowl. Add salad dressing. Mix well and refrigerate for a couple of hours. Serve.

零失敗技巧
Successful Cooking Skills

可以用其他鮮果嗎？

可以配其他鮮果，味道同樣美味。

Can I use other fruits instead?

Yes, you can use any fruit you like. They taste equally great with the dressing.

將香茅莖拍碎和切碎的分別大嗎？

將香茅莖切碎，會令香茅味更加濃郁。

Is there any difference between bruising and finely chopping the lemongrass?

The lemongrass taste would be stronger if you finely chop it.

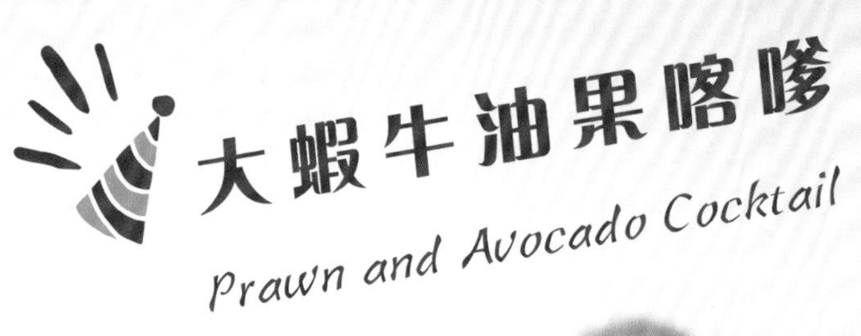

大蝦牛油果喀嗲
Prawn and Avocado Cocktail

材料（4 人份量）

大蝦 4 隻
車厘茄數粒
西芹 1 枝
青瓜粒 1/2 杯
牛油果 1 個
生菜葉數片
檸檬汁少許

灼蝦料

洋葱 1/4 個（切絲）
香葉 2 片

醬料

千島醬 4 湯匙

做法

1. 蝦去頭、去腸，洗淨。
2. 水 1 杯，加入洋葱絲、香葉煮滾，放入蝦灼熟，盛起，攤凍，去殼。
3. 青瓜粒用少許鹽腌片刻；車厘茄切半；西芹撕去筋，切粒。
4. 牛油果去皮去核，切粒，灑下檸檬汁拌勻。
5. 盛器內先放上生菜葉，再放上青瓜粒、牛油果粒和西芹粒，擠上部分千島醬，再放上蝦和車厘茄，加入餘下的千島醬，雪藏片刻，便可以品嘗這美味的沙律了。

Ingredients (Serves 4)

4 prawns
a few cherry tomatoes
1 stalk celery
1/2 cup diced cucumber
1 avocado
a few lettuce leaves
a dash of lemon juice

Ingredients for scalding prawns

1/4 onion (shredded)
2 bay leaves

Sauce

4 tbsp thousand island salad dressing

Method

1. Remove the heads and black veins from prawns. Rinse.
2. Pour 1 cup of water in pot. Add shredded onion and bay leaves and bring to the boil. Put in prawns and scald until done. Remove and leave to cool. Remove the shells.
3. Marinate cucumber with a pinch of salt for a while. Cut cherry tomatoes in half. Tear the hard strings off celery and dice.
4. Peel and stone avocado. Dice. Sprinkle with lemon juice and mix well.
5. Place lettuce into container. Add cucumber, avocado and celery. Squeeze some of the thousand island salad dressing onto the vegetables and fruit. Put in prawns and cherry tomatoes. Add the remaining thousand island salad dressing. Refrigerate for a while. Serve.

零失敗技巧
Successful Cooking Skills

怎樣可去除蝦的腥味？

可用洋葱和香葉去除腥味，因它有增香辟腥的功效。

How do you remove the fishy taste of prawns?

Boil them with onion and bay leaves would make them less fishy. The onion and bay leaves also enhance the aroma of the dish.

牛油果容易變黑，看來不新鮮，哪怎麼辦？

牛油果與檸檬汁調勻，可令牛油果不易變黑，賣相更美。

Avocado tends to blacken very quickly after peeled and sliced. What should I do to keep it bright and green?

Squeeze some lemon juice on the avocado and mix well. The lemon juice will stop it from turning black and the dish will look much better.

柚子沙律

Pomelo Salad

材料（4 人份量）

泰國金柚 1/2 個
腰果 80 克
中蝦 8 隻
乾葱 2 個
炸乾葱 1 湯匙
薄荷葉少許
香葉 2 片
洋葱絲 1/4 個份量

汁料

泰國辣椒膏 2 茶匙
魚露 1 1/2 湯匙
青檸汁 2 湯匙
椰糖 1 /2 湯匙

炸乾葱材料

乾葱數粒
鹽少許

做法

1. 金柚去皮去衣，將果肉撕幼，待用。
2. 蝦去殼去腸，在背部剔一刀，用少許鹽、粟粉拌勻醃一會，沖淨。煮滾 1 杯水，加入洋葱絲、香葉，放入蝦仁灼熟，盛起。
3. 乾葱去衣，橫切薄片。
4. 熱水內下少許鹽，放入腰果浸片刻，盛起。待腰果吹乾後，用油炸至金黃色。
5. 將汁料調勻。
6. 將所有材料（除腰果外）與汁料拌勻，在進食前加入腰果和薄荷葉便可以。

Ingredients (Serves 4)

1/2 Thai pomelo
80 g cashewnuts
8 medium shrimps
2 shallots
1 tbsp deep fried shallots
a few mint leaves
2 bay leaves
1/4 onion, shredded

Sauce

2 tsp Thai chilli paste
1 1/2 tbsp fish sauce
2 tbsp lime juice
1/2 tbsp palm sugar

Ingredients for deep frying shallots

a few shallots
a pinch of salt

Method

1. Peel pomelo and skin. Tear the flesh into small pieces. Set aside.
2. Shell shrimps and remove the veins. Cut along the back. Marinate with a pinch of salt and cornflour for a while. Rinse. Bring 1 cup of water to the boil. Put in onion and bay leaves. Add shrimps and scald. Remove.
3. Skin shallots. Thinly and horizontally slice.
4. Add a pinch of salt to hot water. Put in cashewnuts and soak for a while. Remove. Leave to dry. Deep fry until golden brown.
5. Mix sauce well.
6. Mix all the ingredients (except cashewnuts) with sauce. Add cashewnuts and mint leaves just before serving.

炸乾葱做法

乾葱去衣，洗淨後切片，用少許鹽拌匀醃片刻，擠乾水分，用油炸至乾身和淺黃色，盛起，待涼。

Deep frying shallots

Skin shallots. Wash and slice. Marinate with a pinch of salt for a while. Squeeze out the water. Deep fry until dry and light golden brown. Remove and leave to cool.

零失敗技巧
Successful Cooking Skills

椰糖在哪裏有售？

椰糖在東南亞食品店有售。椰糖味道芳香、獨特，是其他糖類不能代替的。

Where can I get palm sugar?

You can get it from grocery stores specializing in Southeast Asian products. Palm sugar has a unique aroma and it cannot be replaced by other sugars.

怎樣可以令炸乾葱更加香脆？

除了用鹽醃一會，並擠去水分外，可再放在乾毛巾或廚紙上索乾水分，這樣乾葱炸後會格外香脆。

How can I make the deep fried shallot even crispier?

Add a pinch of salt and mix well to draw the moisture out. Then wipe it dry in dry towel or with kitchen paper. It will turn out extra-crispy after fried that way.

三文魚南瓜芝士暖沙律

Warm Salad with Salmon, Pumpkin and Cheese

材料（4 人份量）

南瓜 120 克
羊奶芝士 90 克（feta）
紫洋葱 1/4 個
煙三文魚 100 克

沙律汁

芝麻醬 1 湯匙
檸檬汁 1 1/2 湯匙
橄欖油 2 湯匙
麻油 1 茶匙
黑椒碎少許
鹽 1/2 茶匙

做法

1. 南瓜去部份皮，切件，放入已預熱 190℃的焗爐內焗約 15 分鐘至熟。
2. 紫洋葱切碎，羊奶芝士揑碎。
3. 煙三文魚切片，調勻沙律汁。
4. 紫洋葱、羊奶芝士、煙三文魚和南瓜件放入盛器內，澆上沙律汁即可享用。

Ingredients (Serves 4)

120 g pumpkin
90 g feta cheese
1/4 red onion
100 g smoked salmon

Salad dressing

1 tbsp sesame paste
1 1/2 tbsp lemon juice
2 tbsp olive oil
1 tsp sesame oil
a pinch of ground black pepper
1/2 tsp salt

Method

1. Peel part of the pumpkin. Cut into pieces. Transfer to oven which has been preheated at 190°C. Bake for 15 minutes or until done.
2. Finely chop red onion. Pinch feta cheese into small pieces.
3. Slice salmon. Mix the salad dressing well.
4. Put red onion, feta cheese, salmon and pumpkin into a big bowl. Pour in salad dressing. Serve.

零失敗技巧
Successful Cooking Skills

宜用甚麼種類的南瓜？

宜用日本南瓜，它味道香甜、肉質粉嫩，焗烤後南瓜呈焦糖化非常美味。

What kind of pumpkin should I use for this recipe?

I prefer Japanese pumpkin for its sweetness and starchy texture. It caramelizes nicely after grilled and gives the salad an extra dimension of flavours.

若家中沒有焗爐，怎麼辦？

可以用蒸法代替焗爐烤焗南瓜。

What should I do if I don't have an oven?

You can steam the pumpkin instead of baking it.

這沙律可以預先準備嗎？

可以先處理紫洋葱、羊奶芝士、煙三文魚，調勻沙律汁，臨食前才焗熟南瓜，再撈勻其餘材料，倒下汁料就可享用。

Can I make this salad ahead of time?

You can prepare the red onion and feta cheese, smoked salmon, and make the dressing in advance. Just bake the pumpkin right before you serve. Toss all ingredients together and drizzle with the salad. Serve.

牛蒡沙律

Burdock Salad

材料（4 人份量）

牛蒡 1/2 條
蓮藕 1 小節
雲耳 2 朵
西芹 1 支
糖、鹽各 1/4 茶匙
炒香芝麻 2 茶匙

醋水料

米醋 1 湯匙
水 2 杯
——調勻，分成 2 份

沙律汁

米醋 1 湯匙
乾葱茸 1 茶匙
鹽 1/8 茶匙
黑椒碎少許
芝麻醬 1 茶匙
芝麻碎 1 茶匙
麻油 2 湯匙

做法

1. 刮去牛蒡外皮，切成幼條，浸入 1 份醋水內。
2. 蓮藕去皮，切幼條，浸入另 1 份醋水內。
3. 雲耳浸軟，放入沸水中灼片刻，盛起，切細件。
4. 盛起牛蒡絲，加入各 1/4 茶匙鹽、糖，拌勻醃片刻，擠乾水分。
5. 將蓮藕絲、牛蒡絲、雲耳絲和西芹絲拌勻，澆汁料，灑上芝麻便可以享用。

Ingredients (Serves 4)

1/2 burdock
1 small section lotus root
2 cloud ear
1 sprig celery, shredded
1/4 tsp sugar
1/4 tsp salt
2 tsp stir-fried sesame

Diluted vinegar

1 tbsp rice vinegar
2 cups water
— mixed well and divided into 2 portions

Salad dressing

1 tbsp rice vinegar
1 tsp finely chopped shallot
1/8 tsp salt
a pinch of ground black pepper
1 tsp sesame paste
1 tsp ground sesame
2 tbsp sesame oil

Method

1. Scrape off the outer skin of burdock. Cut into thin strips. Soak in one portion of diluted vinegar.
2. Peel lotus root. Cut into thin strips. Soak in another portion of diluted vinegar.
3. Soak cloud ear until soft. Blanch for a while. Remove. Cut into small pieces.
4. Remove burdock. Add 1/4 tsp of salt and 1/4 tsp of sugar. Mix well and marinate for a while. Squeeze out the water.
5. Mix lotus root, burdock cloud ear and celery. Pour the salad dressing. Sprinkle with sesame and serve.

零失敗技巧
Successful Cooking Skills

牛蒡、蓮藕容易變成鐵繡色，怎麼辦？

牛蒡、蓮藕容易氧化變成鐵繡色，可用醋水浸一會以防變色。

Burdock and lotus root discolour easily after cut. What can I do to prevent that?

Burdock and lotus root turns brown after exposed to air. You can soak them in water with a dash of vinegar for a while to prevent discolouration.

牛蒡、蓮藕可以用大滾水灼一會？

可以的，但不要灼過久，否則牛蒡、蓮藕會不夠爽脆。

Can I blanch burdock and lotus root in boiling water?

Yes, you can. But don't blanch them for too long. Otherwise, they turn soggy instead of crunchy.

吞拿魚沙律船

Cucumbers Stuffed with Tuna Salad

材料（4 人份量）

小青瓜 4 條
吞拿魚 1 罐（95 克）
沙律醬 3 湯匙

調味料

鹽少許
黑椒粉少許

做法

1. 隔去吞拿魚水分，然後再壓碎。
2. 加入沙律醬和調味料拌勻成吞拿魚醬。
3. 青瓜洗淨抹乾，橫切一半，再在中間直切分成兩條，挖去瓜瓤待用。
4. 將適量吞拿魚醬釀在青瓜上即成，想更美觀可綴上裝飾。

Ingredients (Serves 4)

4 small cucumbers
1 can tuna (95 g)
3 tbsp salad dressing

Seasoning

a pinch of salt
a pinch of ground black pepper

Method

1. Drain tuna and then crush.
2. Add salad dressing and seasoning. Mix well to make tuna fillings.
3. Wash cucumbers and wipe dry. Cut in half horizontally. Then cut each half into 2 pieces vertically. Remove the pith. Set aside.
4. Stuff cucumbers with suitable amount of tuna fillings. Put on garnish if desired.

零失敗技巧
Successful Cooking Skills

對挑選青瓜，有甚麼建議？

宜挑選外型較小的青瓜，賣相會更加精緻。

What should I look for when shopping for cucumbers?

Pick those in smaller sizes. That would make the dish look better.

市場上有油浸吞拿魚罐頭和水浸吞拿魚罐頭，有甚麼分別呢？

其實味道的分別不大，油浸吞拿魚的味道會較香和鹹，但我會選用水浸吞拿魚罐頭，因它的脂肪含量較低。

Canned tuna comes in brine or in oil. What are their differences?

They taste the same generally. Canned tuna in oil is richer and saltier. But I prefer canned tuna in brine for lower fat content.

可以預先炮製這個吞拿魚沙律船嗎？

可以分開調勻吞拿魚醬和處理青瓜，放在雪櫃冷藏，食前才將吞拿魚醬釀入青瓜內。

Can I make this dish ahead of time?

You can make the tuna salad and prepare the cucumbers in advance. Keep them in the fridge separately. Just stuff the cucumbers with the tuna salad right before serving.

除了用吞拿魚做餡料外，還可以用甚麼代替？

可以用罐頭蟹肉、牛油果粒做餡料，舀在青瓜船上，再澆上日式芝麻醬，非常美味。

Apart from canned tuna, what else can I use for the filling?

Feel free to use canned crabmeat and diced avocado as the filling. Just spoon the filling on the cucumber and drizzle with Japanese sesame dressing. It tastes divine.

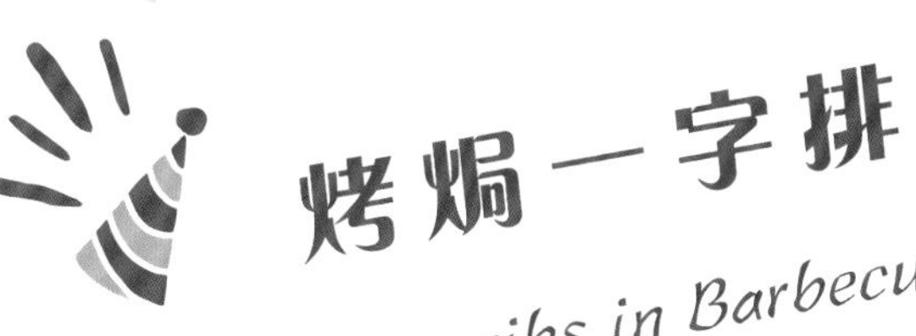

烤焗一字排

Grilled Spareribs in Barbecue Sauce

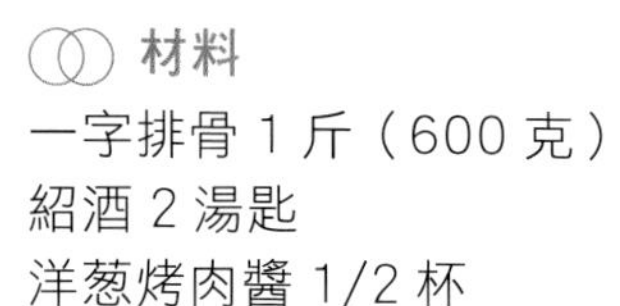

材料

一字排骨 1 斤（600 克）
紹酒 2 湯匙
洋葱烤肉醬 1/2 杯

做法

1. 一字排洗淨，下紹酒及洋葱烤肉醬抹勻，醃 2 小時。
2. 焗盤鋪上錫紙，排上一字排，醬汁留用。
3. 預熱焗爐約 10 分鐘，放入一字排，用 200℃焗 20 分鐘，翻轉排骨，再掃上餘下之洋葱烤肉醬，用 180℃焗 15 分鐘至全熟，趁熱享用。

Ingredients

600 g pork spareribs
2 tbsp Shaoxing wine
1/2 cup onion barbecue sauce

Method

1. Rinse the ribs. Add Shaoxing wine and onion barbecue sauce. Rub evenly. Leave them for 2 hours.
2. Line a baking tray with aluminium foil. Arrange the ribs on the tray.Keep the sauce for later use.
3. Preheat an oven to 200°C for 10 minutes. Bake the ribs for 20 minutes. Flip the ribs upside down. Brush on the remaining onion barbecue sauce. Bake at 180°C for 15 more minutes until fully done. Serve hot.

洋葱烤肉醬

Onion Barbecue Sauce

材料

洋葱 1/2 個（切絲）
蒜肉 8 粒（拍鬆）
黑胡椒碎 1 茶匙

調味汁（拌勻）

蠔油、生抽、紹酒、麥芽糖各 2 湯匙
老抽 2 茶匙

做法

燒熱鑊下油 3 湯匙，下洋葱、蒜肉及黑胡椒碎炒香，加入調味汁煮至濃稠，試味，待涼，入瓶，放於雪櫃可儲存 3 天。

Ingredients

1/2 onion (shredded)
8 cloves garlic (crushed gently)
1 tsp ground black pepper

Seasoning (mixed well)

2 tbsp oyster sauce
2 tbsp light soy sauce
2 tbsp Shaoxing wine
2 tbsp maltose
2 tsp dark soy sauce

Method

Heat a wok and add 3 tbsp of oil. Stir fry onion, garlic and black pepper until fragrant. Add seasoning and cook until thick. Taste it. Leave it to cool. Store in sterilized bottles. It lasts in the fridge for 3 days.

零失敗技巧 Successful Cooking Skills

若醃味時間不足，怎辦？

用叉在排骨上戳小孔，令醃料迅速滲入肉內，毋須擔心時間不足。

I'm running out of time and can't marinate the ribs for long. What should I do?

Just pierce the ribs repeatedly with a fork before marinating. That would make the flavour infuse faster and you don't need to worry about insufficient marinating time.

醃一字排骨是否有先後次序？

宜先下酒才下洋葱烤肉醬，因酒可辟去肉類的羶味。

When you marinate the ribs, do you put in the marinade ingredients in a particular order?

I usually add wine first before putting in the onion barbecue sauce. It's because the wine helps remove the gamey taste of the meat.

香煎牛柳粒

Fried Beef Tenderloin Cubes

材料（4 人份量）

急凍安格斯牛柳粒約 225 克
蒜肉 4 粒（切片）
牛油 1 茶匙

調味料

黑胡椒碎 1 茶匙
海幼鹽 1/4 茶匙
橄欖油 1 湯匙
* 調勻

做法

1. 牛柳粒解凍，抹淨血水，放入調味料拌勻。
2. 平底鑊下油半湯匙，下蒜片炒香，放入牛柳粒用中火煎至表皮呈金黃，最後下牛油拌勻炒香即成。

Ingredients (Serves 4)

225 g frozen Angus beef tenderloin cubes
4 cloves skinned garlic (sliced)
1 tsp butter

Seasoning

1 tsp grated black pepper
1/4 tsp fine sea salt
1 tbsp olive oil
* mixed well

Method

1. Defrost the tenderloin. Wipe the blood off. Mix in the seasoning.
2. Put 1/2 tbsp of oil in a pan. Stir-fry the garlic until fragrant. Put in the tenderloin and fry over medium heat until golden. Add the butter and stir-fry until scented. Serve.

零失敗技巧
Successful Cooking Skills

安格斯牛柳有何特點嗎？
安格斯牛柳肉質脸滑，而且包含濃濃的牛肉香味。

Why do you use Angus beef tenderloin for this recipe?
Angus beef tenderloin is tender and fine in texture, while delivering robust and strong meaty flavours.

最後為何加入牛油略炒？
可品嘗帶牛油香的牛柳粒，且帶微焦，酥香入口！若怕脂肪量高，可省掉牛油。

Why did you add an extra knob of butter at last?
The butter gives the beef a lightly browned crust and a characteristic aroma. For those who are concerned with fat intake, you may skip the knob of butter added at last.

以海幼鹽調味，食味有何特別？
海鹽沒經精鍊的過程，保留了鹽份的原味，能吃出牛柳的原汁原味。

You season the beef with fine sea salt only. How does it enhance the flavours?
Sea salt is not refined and it retains the original taste of sea water. Seasoning the beef with sea salt only lets the authentic taste of the beef come through.

巴東牛肉

Beef Rendang

材料（4 人份量）
急凍牛柳 900 克
紅辣椒 2 隻（切碎）
乾葱頭 6 粒（剁茸）
蒜肉 6 粒（剁茸）
椰汁 1 1/2 杯

香料
南薑 1 小塊（切碎）
香茅 1 枝（切碎）
芫茜粉及小茴粉各 1 茶匙
咖喱粉 1 湯匙
印尼蝦膏 1 茶匙
椰糖少許

調味料
鹽適量

做法
1. 牛柳放於雪櫃的下層自然解凍，飛水，瀝乾水分。
2. 燒滾水 4 杯，下牛柳煮 30 分鐘，盛起待涼，切件，牛柳湯汁 2 杯留用。
3. 燒熱油 1 湯匙，下乾葱茸、蒜茸及紅辣椒碎爆香，加入牛柳件炒勻。
4. 下香料（椰糖除外）、椰汁 3/4 杯及牛肉湯汁 2 杯，煮約 45 分鐘至汁液略收乾。
5. 最後加入餘下的椰汁、椰糖及鹽調味，煮至汁液收乾即成。

Ingredients (Serves 4)
900 g frozen beef tenderloin
2 red chillies (chopped up)
6 shallots (finely chopped)
6 cloves skinned garlic (finely chopped)
1 1/2 cups coconut milk

Spices
1 small piece galangal (chopped)
1 stalk lemongrass (chopped)
1 tsp coriander powder
1 tsp cumin powder
1 tbsp curry powder
1 tsp Indonesian shrimp paste
palm sugar

Seasoning
salt

Method
1. Defrost the beef tenderloin in the lower chamber of the refrigerator. Scald and drain.
2. Bring 4 cups of water to the boil. Cook the beef tenderloin for 30 minutes. Remove and leave to cool down. Cut into pieces. Reserve 2 cups of the beef stock for later use.
3. Heat up 1 tbsp of oil. Stir-fry the shallot, garlic and red chilli until fragrant. Add the beef tenderloin and stir-fry evenly.
4. Put in the spices (except the palm sugar), 3/4 cup of the coconut milk and 2 cups of the beef stock. Cook for about 45 minutes until the sauce nearly dries out.
5. Add the remaining coconut milk. Season with the palm sugar and salt. Cook until the sauce dries out and serve.

零失敗技巧
Successful Cooking Skills

除牛柳外，還可選其他部份嗎？
急凍牛肋條也是不錯之選擇，肉汁香濃有咬口。

Can I use other cuts of beef besides beef tenderloin?

Yes, you can. Frozen beef rib fingers also work well with this recipe as they are juicy and rich with some chewiness.

牛肉湯汁泛着油，如何去掉？
將牛肉湯汁放於雪櫃凝固，可將泛起的脂肪油層去掉，吃起來更健康。

There is a layer of oil over beef sauce. How can I get rid of it?

You can put the sauce in the fridge until the oil solidifies. Then just lift the solid grease off. The sauce will be less fattening and more healthful that way.

加入椰糖後容易黏底，怎辦？
改用慢火，煮時經常翻動，以免焦燶。

The sauce tends to burn easily after I add palm sugar. What should I do?

Cook over low heat and keep stirring it. That would stop it from burning.

越式咖喱雞伴蒜茸包
Vietnamese Style Chicken Curry with Garlic Bread

材料（4 人份量）
雞髀 2 隻（大）
洋葱 1 個（切碎）
馬鈴薯 2 個（大，切件）
紅蘿蔔 1 條（切件）
香茅 2 枝（切碎）
紅辣椒 1 隻（去籽，切碎）
咖喱醬 2 湯匙
蒜茸及乾葱茸各 1 1/2 湯匙
薑茸 2 茶匙
紅辣椒粉 1 茶匙
椰汁半杯

醃料
鹽 1/2 茶匙
胡椒粉少許

調味料
魚露 1 湯匙
鹽 3/4 茶匙
糖 1 茶匙

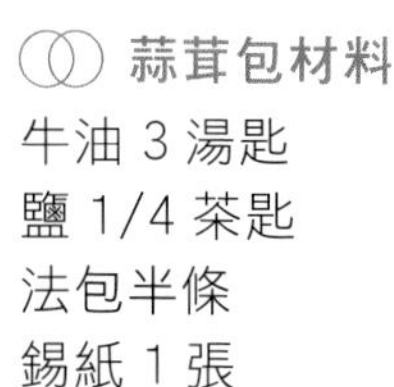

蒜茸包材料

牛油 3 湯匙
鹽 1/4 茶匙
法包半條
錫紙 1 張

做法

1. 雞髀斬件，用醃料拌勻。
2. 塗蒜茸包的牛油放室溫待軟化，再與其他材料拌勻；法包斜切成塊狀。
3. 燒熱少許油，下馬鈴薯略炸至金黃色，盛起，傾出油。
4. 下蒜茸、乾葱茸、薑茸、洋葱、香茅及紅椒碎略炒，加入咖喱醬及紅椒粉拌炒，放入雞件炒勻，下紅蘿蔔及水 1 杯煮 15 分鐘。
5. 加入馬鈴薯再煮 15 分鐘，最後拌入椰汁續煮片刻，下調味料拌勻，即可伴蒜茸包享用。

蒜茸包做法

1. 將牛油蒜茸塗於法包上，排好，用錫紙包好。
2. 放入預熱焗爐，用 190℃焗約 15 分鐘，打開錫紙，蒜茸牛油面向上，再焗片刻即成。

Ingredients (Serves 4)

2 large chicken legs
1 onion (chopped)
2 large potatoes (cut into pieces)
1 carrot (cut into pieces)
2 stalks lemongrass (chopped)
1 red chilli (deseeded; chopped)
2 tbsp curry paste
1 1/2 tbsp finely chopped garlic
1 1/2 tbsp finely chopped shallot
2 tsp finely chopped ginger
1 tsp paprika
1/2 cup coconut milk

Marinade

1/2 tsp salt
ground white pepper

Seasoning

1 tbsp fish sauce
3/4 tsp salt
1 tsp sugar

Ingredients for garlic bread

3 tbsp butter
1/4 tsp salt
1/2 loaf baguette
1 piece aluminum foil

Method

1. Chop the chicken legs into pieces. Mix with the marinade.
2. Rest the butter for making garlic bread at room temperature to soften. Then mix with the other ingredients. Cut the baguette diagonally into pieces.
3. Heat up a little oil. Deep-fry the potatoes slightly until golden. Remove. Pour the oil out.

4. Put in the garlic, shallot, ginger, onion, lemongrass and red chilli. Stir-fry for a moment. Add the curry paste and paprika. Stir-fry. Put in the chicken and stir-fry. Add the carrot and 1 cup of water. Cook for 15 minutes.
5. Put in the potatoes and cook for 15 minutes. Finally stir in the coconut milk and cook for a while. Mix in the seasoning. Serve with the garlic bread.

Method for garlic bread

1. Spread the butter and garlic on the baguette. Arrange the pieces in order. Wrap in aluminum foil.
2. Put in a preheated oven. Bake at 190°C for about 15 minutes. Make the aluminum foil open. Bake for a moment with the garlic butter side up to finish.

零失敗技巧
Successful Cooking Skills

我做的蒜茸包很硬，咬得口也破損，怎辦？

試試我介紹的方法吧！用錫紙包着牛油蒜茸法包焗 15 分鐘，最後揭開錫紙再焗一會，外脆內軟，你會愛上的！

My garlic bread is so hard that it cuts the roof of my mouth. What should I do to improve it?

Try the method I suggested. Wrap the baguette in aluminium foil and bake for 15 minutes. Then remove the aluminium foil and bake for a short while. The garlic bread will then be crispy on the outside and fluffy on the inside. You'd love it.

這道餸很辛辣嗎？

不會！雖然加入紅辣椒及紅椒粉，但由於去掉了紅辣椒籽，辣味指數下降不少。

Is Vietnamese curry very spicy?

No, it's not. Though I use paprika and red chillies, it won't burn your head off because I de-seed the red chillies to take the heat away. No worries.

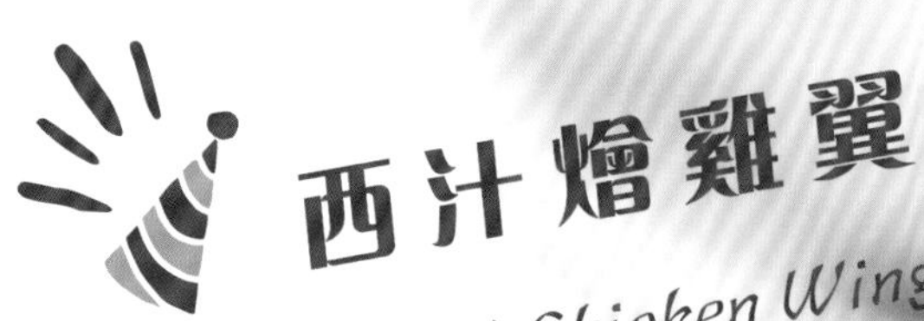

西汁燴雞翼

Braised Chicken Wings in Western Sauce

材料（4 人份量）

雞中翼 600 克
番茜適量

醃料

鹽 1/2 茶匙
生抽 1/2 湯匙
薑汁酒 1 湯匙
糖 3/4 茶匙
生粉 2 茶匙（後下）

調味料

水 6 湯匙
生抽 1 湯匙
喼汁 2 茶匙
茄汁 2 湯匙
紹酒 2 茶匙

做法

1. 雞翼解凍、洗淨，吸乾水分，放入醃料（生粉除外）醃半小時。
2. 熱鑊下油，雞翼灑上生粉拌勻，煎至兩面金黃熟透，瀝乾油分。
3. 煮滾調味料，放入雞翼煮至汁液濃稠，上碟，以番茜裝飾即成。

Ingredients (Serves 4)

600 g chicken mid-joint wings
parsley

Marinade

1/2 tsp salt
1/2 tbsp light soy sauce
1 tbsp ginger juice wine
3/4 tsp sugar
2 tsp caltrop starch (for later use)

Seasoning

6 tbsp water
1 tbsp light soy sauce
2 tsp Worcestershire sauce
2 tbsp ketchup
2 tsp Shaoxing wine

Method

1. Defrost the chicken wings. Rinse and wipe dry. Mix in the marinade (except the caltrop starch) and rest for 1/2 hour.
2. Add oil in a heated wok. Dust the chicken wings with the caltrop starch and mix well. Fry until both sides turn golden and are fully cooked. Drain.
3. Bring the seasoning to the boil. Add the chicken wings and cook until the sauce reduces. Arrange on the plate. Decorate with the parsley. Serve.

零失敗技巧
Successful Cooking Skills

為甚麼生粉最後才拌入？
煎雞翼前才拌入生粉，令雞翼帶香脆的口感，百吃不厭！

Why did you stir in the caltrop starch at last?
Stir in the caltrop starch right before you pan-fry the chicken wings. That's the secret trick to crispy skin on the chicken wings. You can never get enough of it.

蝦堡

Shrimp Burgers

材料（4 人份量）
方包 8 片
千島醬 3 湯匙
牛油 2 茶匙
蝦肉 300 克
麵包糠 4 湯匙

調味料
鹽 1/3 茶匙
胡椒粉少許
生粉 1 茶匙

做法
1. 蝦肉洗淨，用布索乾水分。用刀拍爛，改用刀背剁片刻。
2. 將蝦肉放於深碗中，加入調味料拌勻，再攪至起膠，蝦膠放入雪櫃冷藏片刻。
3. 蝦膠分成兩份，每份搓圓，按扁至直徑與方包相若，沾上麵包糠成蝦餅。
4. 熱鑊下少許油，用慢火將蝦餅煎至熟及兩面金黃色，盛起。
5. 方包塗上牛油和千島醬，中間夾上蝦餅，每份切成兩件便可以享用。

Ingredients (Serves 4)
8 pieces bread
3 tbsp thousand island salad dressing
2 tsp butter
300 g shelled shrimps
4 tbsp breadcrumbs

Seasoning
1/3 tsp salt
a pinch of pepper
1 tsp caltrop starch

Method
1. Wash shrimps and use a piece of cloth to absorb the moisture. Pound with a knife. Chop for a while with the back of the knife.
2. Place shrimps in a deep bowl. Mix with seasoning. Stir until sticky. Refrigerate for a while.
3. Divide shrimp paste into 2 portions. Shape each into a ball. Flatten until the diameter is more or less the same as that of bread. Coat with breadcrumbs to make shrimp patty.
4. Heat wok and add a little oil. Saute shrimp paste over low heat until done and both sides are golden brown. Remove.
5. Spread bread with butter and thousand island salad dressing. Sandwich 2 pieces of bread with prawn patty. Cut each burger in half. Serve.

零失敗技巧
Successful Cooking Skills

炮製蝦膠有甚麼要注意之處？

最重要是索乾蝦的水分，以免蝦肉含高水分而變「霉」。有個簡單的辦法，是將蝦放在乾淨的乾毛巾上，捲起，放入雪櫃冷藏約 1 小時，就可索乾水分。

What are the secret tricks to bouncy and tasty minced shrimp?

Make sure you wipe the shrimps completely dry before chopping. Otherwise, the extra moisture would make the minced shrimp soggy instead of bouncy after cooked. One way to do it, is to put the shelled shrimps on a clean dry towel and roll it up. Refrigerate for 1 hour to draw moisture out.

炮製蝦膠要用游水鮮蝦嗎？

可以用冰鮮蝦，蝦味會較濃郁。

Should I use live shrimps to make minced shrimp?

You may use frozen shrimps as they tend to have stronger seafood flavours.

泰式香辣雞翼

Thai Spicy Chicken Wings

材料（4 人份量）
雞中翼 600 克
香茅碎、乾葱碎、蒜茸各 2 湯匙
雞蛋 1 個（拂勻）
生粉、自發粉各 2 湯匙

醃料
鹽 1/2 茶匙
魚露 1 茶匙
胡椒粉少許

汁料
泰式雞醬 2 湯匙
橙汁 4 湯匙
橙 1/2 個（切碎）
泰式椰糖 1 湯匙

做法
1. 雞中翼洗淨，抹乾水分，加入醃料拌勻。
2. 將香茅碎、乾葱碎、蒜茸舂爛，用布包好，將汁液榨出；汁液與雞翼一起拌勻，醃約 3 小時。
3. 將汁料煮約 10 分鐘至濃稠，隔渣，待用。
4. 雞翼與蛋汁調勻，拌入生粉、自發粉，雞翼放入熱油內炸至金黃色，盛起，蘸汁料享用。

Ingredients (Serves 4)
600 g mid-joint chicken wings
2 tbsp finely chopped lemongrass
2 tbsp finely chopped shallot
2 tbsp finely chopped garlic
whisked egg
2 tbsp caltrop starch
2 tbsp self-raising flour

Marinade
1/2 tsp salt
1 tsp fish sauce
a pinch of pepper

Sauce
2 tbsp sweet chilli sauce
4 tbsp orange juice
1/2 orange (finely chopped)
1 tbsp Thai coconut sugar

Method
1. Wash chicken wings and wipe dry. Mix with marinade.
2. Pound lemongrass, shallot and garlic. Wrap in a piece of cloth. Squeeze out the juice. Mix chicken wings with the juice. Marinate for 3 hours.
3. Cook the sauce for 10 minutes or until thick. Strain and set aside.
4. Mix chicken wings and whisked egg. Stir in caltrop starch and self-raising flour. Deep fry chicken wings in hot oil until golden brown. Remove. Dip in sauce and serve.

零失敗技巧
Successful Cooking Skills

為甚麼要將雞翼先拌入蛋汁，才再拌入粉類？

雞翼拌入蛋汁的原因是，可讓生粉和自發粉緊黏雞翼；撲上粉類的雞翼宜即炸，否則粉料會變濕，這樣雞翼外皮的香脆度會減低。

Why do you dunk the chicken wings in whisked egg first, before coating them in dry ingredients?

The whisked egg helps bind the caltrop starch and self-raising flour onto the wings. After you coat the wings in dry ingredients, you should deep-fry them right away. Otherwise, the starch and flour will pick up the moisture from the wings and they will turn stiff and dense instead of light and crispy after fried.

這個泰式香辣雞翼可以預先炮製嗎？

可以的，若想雞翼更加香脆，臨吃前翻炸一會即可。

Can I make this dish ahead of time?

Yes, you can. If you want the wings to be extra-crispy, just deep-fry them once more right before serving.

為甚麼雞翼會不太入味？

在醃雞翼前應抹乾水分，這樣醃料才容易滲進肉內。

How come my wings don't pick up the seasoning too well?

You should wipe the wings completely dry before marinating, so that the moisture won't dilute the marinade. It also helps the marinade to infuse into the flesh.

番茄魔鬼蛋

Deviled Eggs with Tomatoes

材料（8 人份量）

雞蛋 4 個
番茄膏 1 湯匙
油浸番茄乾 2 件
沙律醬 2 湯匙

做法

1. 雞蛋放入煲內，加入過面水，水煮滾後 5 分鐘關火，再焗 2 分鐘，盛起再浸凍水。
2. 雞蛋去殼，切半。
3. 取出蛋黃，壓爛。
4. 油浸番茄乾切成幼粒，與其他材料加入蛋黃中拌勻成餡料。
5. 將餡料舀進擠花袋中，然後擠在蛋白上便成。

Ingredients (Serves 8)

4 eggs
1 tbsp tomato paste
2 sundried tomatoes soaked in oil
2 tbsp salad dressing

Method

1. Put eggs in pot. Add water until eggs are immersed. Bring water to the boil and simmer for 5 minutes. Turn off the stove and cover for 2 minutes. Remove and soak in cold water.
2. Shell eggs and cut in half.
3. Remove egg yolks and crush.
4. Finely dice dried tomatoes. Mix with the remaining ingredients and egg yolks to make fillings.
5. Put the fillings in forcing bag. Squeeze the fillings onto egg yolks. Serve.

零失敗技巧
Successful Cooking Skills

有甚麼方法令雞蛋更易剝殼？
雞蛋烚熟後浸凍水，會更容易剝去蛋殼。

Is there any way to make the hard-boiled eggs easier to shell?
Soak them in cold water right after boiling them. The thermal shock would make them easier to shell.

番茄醃肉撻

Tomato Tart with Bacon

撻皮（4 吋批撻模 4 個）

麵粉 150 克
鹽少許
牛油 90 克（切成細粒）
蛋黃 1 個
冰水約 2 至 3 湯匙

餡料

番茄乾 40 克
加拿大醃肉 75 克
雞蛋 2 個
蛋黃 1 個
忌廉 200 克

調味料

鹽 1/2 茶匙
黑椒粉少許

撻皮做法

1. 麵粉及鹽同篩勻，加入牛油粒，用手指搓成麵包糠狀。
2. 加入蛋黃拌勻，再注入冰水略搓，靜置 20 分鐘。
3. 將麵糰擀薄至撻模大小，按入模內，用叉子刺孔。
4. 撻內鋪上烤焗用的豆，放入預熱焗爐，以 190℃焗約 10 分鐘，備用。

餡料做法

1. 醃肉及番茄乾分別切碎。
2. 雞蛋、蛋黃及忌廉拂勻，加入調味

綜合做法

1. 將餡料均勻地分放於略烘焗的撻皮上。
2. 放入預熱焗爐，以 190℃焗約 25 分鐘即成。

Tart crust (Makes four 4-inch tarts)

150 g flour
salt
90 g butter, diced
1 egg yolk
2 to 3 tbsp iced water

Filling

40 g sun-dried tomatoes in oil
75 g Canadian cured bacon
2 eggs
1 egg yolk
200 g whipping cream

Seasoning

1/2 tsp salt
ground black pepper

Crust

1. Sieve flour and salt together. Dice the butter and put them in the dry ingredients. Rub with the tips of your fingers until crumbly.
2. Add egg yolk and mix well. Add iced water and knead into dough. Leave it to rest for 20 minutes.
3. Roll the dough out with a rolling pin according to the size of the tart mould. Press the dough into the tart mould. Pierce evenly with a fork.
4. Fill with the baking beans in the tart mould. Bake in a preheated oven at 190°C for 10 minutes. Set aside.

Filling

1. Drain the sun-dried tomatoes. Finely chop the bacon and sun-dried tomatoes.
2. Whisk the eggs, egg yolk and whipping cream. Add seasoning and stir well. Stir in the rest of the ingredients.

Assembly

1. Put the filling evenly on each baked tart crust.
2. Bake in a preheated oven at 190°C for 25 minutes. Serve.

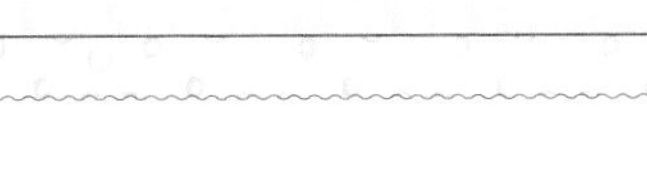

零失敗技巧
Successful Cooking Skills

為甚麼撻皮在烘焙會拱起？

因為在撻皮和撻模之間有空氣，受熱時會膨脹；所以要在撻皮按入模內後，用叉子在撻皮上刺孔，讓空氣溢出。

Why does the tart crust puff up when baked?

The air trapped between the tart tin and the pastry would expand when heated up in the oven. Thus, after you press the pastry into the tart tin, you should pierce it repeatedly with a fork. That would release the air and prevent the tart crust from puffing up.

將麵粉、鹽和牛油用手指捏成麵包糠狀，有何秘訣？

要將牛油雪硬才切成幼粒，與麵粉撈勻，才容易捏成麵包糠狀。用手指頭的原因是它較冰涼，令牛油在捏時保持冰凍。

What are the tricks in rubbing the flour, salt and butter together with fingertips?

Make sure the butter is chilled and hard before dicing it. Then mix with flour and rub them together with your fingertips into in crumbly mixture. Use your fingertips because they are cooler than the palm so that the butter won't melt in the process.

金粟菜肉煎餃
Sautéed Corn, Vegetable and
Pork Dumplings

材料（可製 24 隻餃子）

絞豬肉 250 克
粟米粒 1/3 杯
椰菜 100 克
冬菇數朵
木耳 1 朵
薄餃子皮 24 張

調味料

鹽 3/4 茶匙
生抽 1 湯匙
糖 1/2 茶匙
胡椒粉少許
酒 1/2 茶匙
粟粉 1 茶匙

做法

1. 椰菜切小塊後，與半茶匙鹽拌勻，半小時後擠去水分；冬菇浸軟切幼粒，木耳浸軟切碎。
2. 絞豬肉先加入鹽、生抽，循一方向攪至起膠。
3. 繼而加入餘下的調味料再攪拌，最後加入冬菇粒、木耳碎、粟米粒、椰菜拌勻成餡料。
4. 將餃子皮放於手上，舀上適量餡料；在餃子皮邊塗少許水，對合，用兩拇指、食指來夾實，再向上推成餃子形。
5. 燒熱少許油，放上餃子略煎，然後加半杯水，加蓋煮至水分收乾，再略煎片刻便可享用。

零失敗技巧
Successful Cooking Skills

為甚麼要用鹽醃椰菜？

可以將椰菜的水分溢出，令餡料乾爽，煎時不會有水分弄濕皮料，令餃子底部保持香脆。

Why do you marinate the cabbage with salt?

That helps draw the moisture out of the cabbage so that the filling won't be too wet. Otherwise, the moisture would make the skin wet when you fry the dumplings and the dumplings won't be crispy on the bottom.

椰菜已有足夠的鹹味，為甚麼豬肉還要加鹽？

豬肉加入鹽，不單只有調味作用，還可在攪拌時讓豬肉容易起膠。

You have seasoned the cabbage with salt already. Why do you add more salt to the pork?

The salt in the pork is not solely for seasoning. The salt helps the ground pork turn sticky more easily when stirred.

Ingredients

(24 dumplings can be made)

250 g minced pork
1/3 cup whole kernel corn
100 g cabbage
a few dried black mushrooms
1 wood ear
24 sheets thin dumpling wrappers

Seasoning

3/4 tsp salt
1 tbsp light soy sauce
1/2 tsp sugar
a pinch of pepper
1/2 tsp wine
1 tsp cornflour

Method

1. Cut cabbage into small pieces. Mix with 1/2 tsp of salt and leave for half an hour. Squeeze out the water. Soak dried black mushrooms until soft and finely dice. Soak wood ear until soft and finely chop.
2. Add salt and light soy sauce to pork. Stir in one direction until sticky.
3. Then add the remaining seasoning and stir well. Put in mushrooms, wood ear, corn and cabbage. Mix well to make fillings.
4. Place a dumpling wrapper on the palm. Put suitable amount of fillings on top. Brush the edge of the wrapper with a little water. Fold and seal with the two thumbs and index fingers. Push upward to make a dumpling.
5. Heat a little of oil. Sauté dumplings for a short while. Pour in 1/2 cup of water. Cover and cook until the water dries up. Sauté for a while and serve.

簡易麵包薄餅

Simple Bread Pizzas

材料（10 件麵包薄餅）

法國麵包 1 條
水牛芝士 (mozzarella) 120 克
茄膏或意粉醬 4 湯匙
沙樂美腸、菠蘿粒、蘑菇片各少許
粟米粒 2 湯匙
青椒絲少許
牛油少許

做法

1. 水牛芝士刨碎。
2. 法國麵包斜切成薄片。
3. 包上塗牛油，再抹上茄膏或意粉醬，放上芝士碎和其他材料，再灑上芝士碎，放入已預熱 190℃的焗爐內焗 15 分鐘即成。

Ingredients

(10 bread pizzas can be made)

1 French baguette
120 g mozzarella cheese
4 tbsp tomato paste or spaghetti sauce
a few salami sausage slices
a little pineapple dice
a few sliced button mushrooms
2 tbsp whole kernel corn
a few shredded green bell pepper
a little butter

Method

1. Shave mozzarella cheese into small pieces.
2. Cut baguette into thin pieces diagonally.
3. Spread baguette with butter. Brush with tomato paste or spaghetti sauce. Place some of the cheese and the other ingredients on baguette. Sprinkle with the remaining cheese. Transfer to an oven which has been preheated at 190°C. Bake for 15 minutes. Serve.

零失敗技巧
Successful Cooking Skills

食材可以替代嗎？

可用任何材料代替沙樂美腸、菠蘿、蘑菇片、粟米粒或青椒絲，但一定不可刪去牛油、茄膏或意粉醬和水牛芝士，它們是麵包薄餅美味的關鍵。

Can I replace with topping with other ingredients?

Yes, you can. Feel free to use salami, pineapple, slice mushrooms, corn kernels or shredded bell pepper. However, you can't omit the butter, tomato paste (or pasta sauce) and the mozzarella because they are the soul of a delicious pizza.

為甚麼要塗上牛油？

麵包塗上牛油，可以防止麵包因食材影響而變濕。

Why do you spread butter on the pizza?

Spreading butter on bread actually stops the bread from picking up the moisture from the topping.

墨西哥芝士紅腰豆薄餅
Kidney Bean and Cheese Burritos

材料（可製 4 件薄餅）

墨西哥薄餅 4 張
火腿 100 克
橄欖油 2 茶匙
青椒、紅洋葱 各 1 個（小）
茄膏 4 湯匙
罐頭紅腰豆 1/2 罐
車打芝士 100 克

做法

1. 火腿切碎，用橄欖油炒片刻，盛起。
2. 青椒去蒂、去籽；洋葱去衣，切碎。芝士刨碎。
3. 焗爐調至 190℃，預熱 10 分鐘。
4. 每張墨西哥薄餅塗上茄膏，將火腿碎、洋葱、青椒、紅腰豆分別放於薄餅上，灑上芝士碎。
5. 放入已預熱的焗爐內焗約 10 分鐘，至芝士溶化和有少許金黃色。

Ingredients (Makes 4 burritos)

4 flour tortillas
100 g cooked ham
2 tsp olive oil
1 green bell pepper
1 small red onion
4 tbsp tomato paste
1/2 can kidney beans
100 g cheddar cheese

Method

1. Chop the ham. Stir fry ham in olive oil. Set aside.
2. Cut off the stem of the green bell pepper. Seed it. Set aside. Peel the onion and chop it. Grate the cheese.
3. Preheat an oven to 190°C for 10 minutes.
4. Spread tomato paste over each tortilla. Arrange ham, onion, green bell pepper and red kidney beans over each tortilla. Sprinkle with cheese on top.
5. Bake in a preheated oven for 10 minutes until cheese melts and lightly browned. Serve.

零失敗技巧
Successful Cooking Skills

待芝士焗至溶化時，餅邊又會焦燶，怎麼辦？
可以將錫紙剪一個大孔，覆在薄餅上，只露出餡料，就可避免這情況。

When the cheese melts, the edge of the pizza is charred already. What should I do?
You can cut a hole at the centre of a sheet of aluminium foil. Put it over the pizza exposing only the topping. In that case, the cheese will melt nicely without the edge of the pizza charring.

可以用其他芝士代替嗎？
可以用巴馬臣芝士代替。

Can I use other types of cheese instead?
Yes, you can use grated parmesan instead.

芝蛋漢堡

Cheese and Egg Burgers

材料（8 件漢堡）

意大利香草包 1 個
雞蛋 2 個
番茄 2 個（切片）
車打芝士 2 片
牛油少許

調味料

鹽、黑椒各少許

做法

1. 蛋拂勻，加入調味料，用少許油炒成滑蛋。
2. 意大利香草包橫切成兩片，塗上牛油。
3. 將滑蛋放於底層包上，再放上番茄片、芝士，蓋上面層的麵包，放入已預熱 180℃的焗爐內焗約 10 分鐘，再切成小件。

Ingredients (8 burgers can be made)

1 focaccia
2 eggs
2 tomatoes (sliced)
2 pieces cheddar cheese
a little butter

Seasoning

a pinch of salt
a pinch of black pepper

Method

1. Whisk eggs. Mix with seasoning. Stir fry with a little oil until eggs are of medium consistency.
2. Cut bread horizontally into two pieces.
3. Put eggs on the lower piece of bread. Add tomato slices and cheese. Top with the other piece of bread. Transfer to oven which has been preheated at 180°C. Bake for 10 minutes. Cut into small pieces.

零失敗技巧
Successful Cooking Skills

如家中沒有焗爐怎麼辦？

可以用平底鑊代替。平底鑊內不用下油，用小火燒熱鑊，放入已夾進蛋、番茄片、芝士的麵包，蓋上鑊蓋，麵包底面烘一會就成。

What should I do if I don't have an oven?

You can use a pan instead. Heat a dry pan over low heat without any oil. Put in the focaccia with eggs, tomato and cheese stuffed in. Cover the lid and toast the bottom of the bun briefly.

在餡料方面可以有其他變化嗎？

如不用意大利香草包，可以在超市購買現成的多士，塗上忌廉芝士，放煙三文魚，飾上刁草，就是很得體的餐前小食。

Can I use other filling instead?

If you don't use a focaccia, you can get sandwich bread from supermarkets. Just spread some cream cheese on and arrange smoked salmon. Garnish with fresh dill. Voila. You have an easy and sumptuous pre-meal snacks ready.

鷹嘴豆醬配比得包

Garbanzo Chickpea Paste with Pitta Bread

材料 (6 人份量)

鷹嘴豆 1 罐（約 425 克）
比得包 2 件
橄欖油 1 湯匙

調味料

蒜茸 1 湯匙
芝麻醬 5 湯匙
檸檬汁 4 湯匙
水 1/4 杯
橄欖油 2 湯匙
鹽 1/4 茶匙
糖 1 茶匙

做法

1. 鷹嘴豆隔去水分，放入攪拌機內，加入調味料，攪爛。
2. 比得包剪成小件，掃上橄欖油，用多士焗爐烘片刻，蘸鷹嘴豆醬享用。

Ingredients (Serves 6)

1 can Garbanzo chickpeas (425g)
2 pieces pitta bread
1 tbsp olive oil

Seasoning

1 tbsp finely chopped garlic
5 tbsp sesame paste
4 tbsp lemon juice
1/4 cup water
2 tbsp olive oil
1/4 tsp salt
1 tsp sugar

Method

1. Drain Garbanzo chickpeas. Put in food processor. Add seasoning and process.
2. Cut pitta bread into small pieces. Brush with olive oil. Toast in a toaster for a while. Dip in chickpea paste.

零失敗技巧
Successful cooking skills

怎樣可令鷹嘴豆質感更加幼滑？
將鷹嘴豆分批倒進攪拌機內攪拌，並用幼網篩過濾，效果會更幼滑。
How can I make the chickpea puree creamier?
When you puree the chickpeas in a blender, do it in small batches. Then press the puree through a fine wire mesh for finer texture.

為甚麼比得包要掃上橄欖油？
因為在烘焙時，令比得包香脆些和不會太乾。
Why do you brush olive oil on the pita bread?
The olive oil will make the bread crispier and less dry when baked.

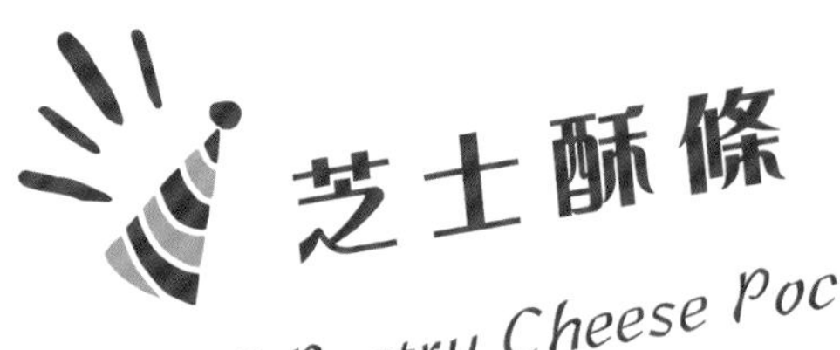

芝士酥條

Puff Pastry Cheese Pockets

材料（可製 15 件）

急凍酥皮 1 包
雞蛋 1 個
麵粉 2 湯匙

餡料

牛油 30 克
麵粉 30 克
牛奶 250 毫升
蛋黃 1 個
edam 芝士 100 克
鹽 1/2 茶匙
黑椒粉少許

做法

1. 芝士刨碎。
2. 牛油與麵粉放煲內，以慢火炒勻，逐少逐少加入牛奶，煮成糊狀，下鹽、黑椒粉調味，攤凍，加入蛋黃和芝士碎拌勻。
3. 急凍酥皮從冰格取出，解凍約 1 小時，酥皮分成兩份。灑麵粉於工作枱上，將酥皮擀薄成 3 毫米厚、24 厘米 x24 厘米的正方形。
4. 每片再裁成三條長條，待用。
5. 每一長條酥皮再剔出 3 厘米，留作做各式形狀的餅面裝飾。
6. 焗爐調至 200℃，預熱 10 分鐘。
7. 將一條長條酥皮放工作枱，等距離放入 5 份適量餡。
8. 在餡的周圍掃上蛋汁，覆上另一條長條酥皮，在餡與餡之間切開，按實周邊。酥皮面掃上蛋液，放上餅面裝飾，再掃蛋汁。
9. 放入已預熱的焗爐內焗約 20 分鐘即成。

Ingredients

(Makes 15 pastries)

1 pack frozen puff pastry
1 egg
2 tbsp flour

Filling

30 g butter
30 g flour
250 ml milk
1 egg yolk
100 g edam cheese
1/2 tsp salt
ground black pepper

Method

1. Grate the cheese.
2. To make the filling, put butter and flour into a pot. Stir over low heat. Slowly pour in milk while stirring until it thickens up. Season with salt and black pepper. Leave it to cool. Add egg yolk and grated cheese. Mix well.
3. Thaw the puff pastry at room temperature for 1 hour. Cut the puff pastry into halves. Sprinkle with flour on the counter. Roll each piece out into a 24-cm square, about 3 mm thick.
4. Cut each square into three 24 cm x 8 cm strips. Set aside.
5. Cut 3 cm of pastry off each 24 cm x 8 cm strip along the length. These narrow strips are to be modelled into different shapes as decorations.
6. Preheat an oven to 200˚C for 10 minutes.

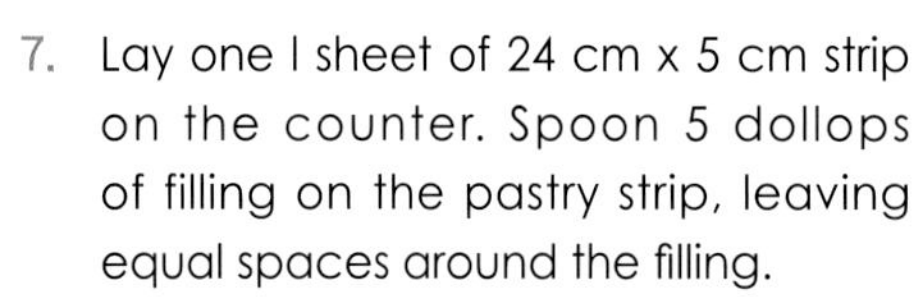

7. Lay one I sheet of 24 cm x 5 cm strip on the counter. Spoon 5 dollops of filling on the pastry strip, leaving equal spaces around the filling.
8. Brush egg wash around the filling. Place another sheet of 24 cm x 5 cm strip over it. Cut the pastry between the filling. Press firmly along the rim. Brush egg wash over the pastry. Garnish with narrow strips. Brush egg wash again.
9. Bake in a preheated oven for 20 minutes.

零失敗技巧
Successful Cooking Skills

可買一片片現成的急凍酥皮嗎？

可以，但它的鬆脆度會稍遜。

Can I use store-bought frozen puff pastry instead?

Yes, you can. But store-bought ones aren't as fluffy and puffy as those made from scratch.

芝士酥條可以預先炮製，吃前才烘焙嗎？

可以，但一定要放進雪櫃貯存，並在烘焙前才掃上蛋汁。

Can I make the puff pastry cheese sticks ahead of time and bake them right before serving?

Yes, you can. Just make sure you refrigerate them. Brush on egg wash right before you bake them.

香葱芝士球

Spring Onion Cheese Balls

材料（可製 16 個芝士球）

水 150 克
麵粉 120 克
牛油 60 克
gouda 芝士 120 克
火腿 2 片（切碎）
雞蛋 3 個
葱粒 3 湯匙
橄欖油 1 茶匙
鹽 1/3 茶匙
黑椒粉少許

做法

1. 用橄欖油炒火腿碎一會；芝士刨碎。
2. 以 200℃爐溫預熱焗爐。
3. 水煮滾後，加入牛油煮溶，下麵粉拌勻，取出攤凍。
4. 雞蛋逐個加入麵糊內打勻。
5. 加入葱花、火腿、芝士碎、鹽、黑椒粉拌勻。
6. 將約 1 湯匙的芝士麵糊舀在焗盤上，每個要相隔約 3 厘米。
7. 放入已預熱 200℃的焗爐內，焗約 20 分鐘即成。

Ingredients

(makes 16 cheese balls)

150 g water
120 g flour
60 g butter
120 g gouda cheese
2 slices ham
3 eggs
3 tbsp diced spring onion
1 tsp olive oil
1/3 tsp salt
ground black pepper

Method

1. Chop the ham finely. Stir fry in a little olive oil briefly. Grate the gouda cheese.
2. Preheat an oven to 200°C.
3. Bring water to the boil in a pot. Put in butter and cook until it melts. Add flour and stir well. Leave it to cool.
4. Crack the eggs and add one egg at a time to the batter. Whisk well after each addition.
5. Add spring onion, ham, grated cheese, salt and black pepper. Stir well.
6. Scoop out about 1 tbsp of batter. Roll it round. Arrange on a lined baking tray. Leave at least 3 cm around each cheese ball.
7. Bake in a preheated oven at 200°C for about 20 minutes. Serve.

零失敗技巧
Successful Cooking Skills

為甚麼雞蛋要逐隻加入麵糊內？

雞蛋是要逐隻加入打透，才可以加第二隻，這樣質感才會鬆軟。

Why do you add one egg at a time into the batter?

I beat the batter until well incorporated after adding each egg. That's how you end up with a fluffy texture.

可以預先準備芝士麵糊放在雪櫃冷藏，臨食前才烘焙嗎？

不可以，要準備完成即烘焙，否則芝士球不會有鬆軟的質感。

Can I make the cheese batter ahead of time and refrigerate it?

No, you can't. You must make the batter and bake it fresh. Otherwise, the cheese balls won't be fluffy.

迷你熱狗多士

Mini Hotdog Toasts

◎ 材料（可製 12 件）

厚方包 2 件
迷你芝士腸 12 條
牛油 1 湯匙
青瓜片適量
茄汁、芥末醬各 1 湯匙

◎ 做法

1. 厚方包去邊，每塊切成 6 件 中間切成 V 狀。
2. 放入焗爐烘片刻至微黃色。
3. 每件麵包塗上少許牛油。
4. 迷你芝士腸放入焗爐烘一會或同沸水內淥片刻，盛起。
5. 將芝士腸放於多士的中間，青瓜片放側，擠上茄汁和芥末即可。

Ingredients

(12 hot dogs can be made)

2 thick pieces bread
12 mini cheese sausages
1 tbsp butter
a few cucumber slices
1 tbsp ketchup
1 tbsp mustard

Method

1. Cut the crusts off bread. Cut each piece into 6 small pieces. Cut the centre into a V shape.
2. Put bread in toaster and toast until light brown.
3. Spread each piece of bread with a little butter.
4. Scald or bake mini cheese sausages for a while. Remove.
5. Put a mini cheese sausage in the centre of each toast. Place cucumber on the side. Squeeze ketchup and mustard on top. Serve.

零失敗技巧
Successful Cooking Skills

焗或淥芝士腸時有甚麼要注意？
芝士腸不要焗或淥過久，否則腸身會爆裂流出芝士。

When I bake or blanch the cheese sausages, is there anything that needs my attention?

Do not bake or blanch them for too long. Otherwise, the sausage may burst and the cheese may leak.

可以用其他醬料嗎？
將第戎芥末醬和蜂蜜調勻，再澆在香腸上也非常美味。

Can I use other condiments on the sausage?

You can make honey mustard by mixing Dijon mustard and honey. Just dribble on the sausage and dig in.

卡邦尼意粉
Carbonara Spaghetti

材料（4 人份量）

意大利粉 150 克
煙肉 100 克
蛋黃 2 個
淡忌廉 60 毫升
巴馬臣芝士粉 3 湯匙
橄欖油 1 湯匙
鹽 1/2 茶匙

做法

1. 煮滾一鍋水，加入少許鹽、橄欖油，再加入意粉煮約 7 分鐘，盛起。
2. 煙肉切薄片，蛋黃拂勻
3. 燒熱 1 湯匙橄欖油，加入煙肉碎，炒至煙肉散發香氣。
4. 注入忌廉、鹽，兜勻。加入意大利粉和芝士粉，兜勻。
5. 最後加入蛋黃，迅速推熟，上碟。

Ingredients (Serves 4)

150 g spaghetti
100 g bacon
2 egg yolks
60 ml whipped cream
3 tbsp grated parmesan cheese
1 tbsp olive oil
1/2 tsp salt

Method

1. Bring a pot of water to the boil. Add a pinch of salt and a dash of olive oil. Put in spaghetti and cook for 7 minutes. Remove.
2. Thinly slice bacon. Whisk egg yolks.
3. Heat 1 tbsp of olive oil. Add bacon and stir fry until fragrant.
4. Pour in whipped cream and salt. Mix well. Add spaghetti and parmesan cheese. Mix well.
5. Finally add egg yolks. Stir quickly until done. Transfer to plate. Serve.

零失敗技巧
Successful Cooking Skills

為甚麼蛋黃要最後才加入？

最後才加入蛋黃，可令汁液更幼滑，因為如一早下蛋黃，會令蛋黃煮得過火，汁液會有微粒。

Why do you add the egg yolk at last?

The egg yolk is added last to keep the sauce creamy and silky. If it's added early on, the egg yolk will be overcooked and the sauce will be lumpy.

煮意大利粉需要過冷河嗎？

勿將意大利粉過冷河，因會洗去麵條上的顆粒，令意大利粉掛不上汁液。

Should I rinse the pasta in cold water after blanching it?

No. You should not rinse pasta in cold water. That would rinse off the starch on the pasta and the sauce won't cling to it.

鰻魚蕎麥麵卷

Eel and Soba Sushi Rolls

材料（32 件份量）

蕎麥麵 150 克
菠菜 150 克
雞蛋 2 個
燒鰻魚 1 條
紫菜 4 張
鰻魚汁或萬用汁適量

用具

壽司蓆

做法

1. 蕎麥麵放入沸水煮約 5 分鐘，用水過冷河，盛起。
2. 菠菜洗淨，用沸水燙熟，過冷河待用。
3. 雞蛋拂勻煎成厚蛋皮，切成長條。
4. 鰻魚切長條。
5. 紫菜放於壽司蓆上，然後放上適量蕎麥麵，再放上菠菜、鰻魚、蛋，捲成壽司。
6. 橫切成小件，伴以萬用汁或少許鰻魚汁享用。

Ingredients (32 rolls can be made)

150 g soba
150 g spinach
2 eggs
1 grilled eel
4 pieces laver
some unagi sauce or concentrated dashi

Utensil

sushi bamboo mat

Method

1. Blanch soba for 5 minutes. Rinse and remove.
2. Wash spinach. Blanch, rinse and set aside.
3. Whisk eggs. Sauté to make thick egg sheet. Cut into long strips.
4. Cut eel into long strips.
5. Place laver on sushi bamboo mat. Put on suitable amount of soba, then spinach, eel and egg sheet. Roll into sushi.
6. Cut sushi into small pieces. Serve with concentrated dashi or unagi sauce.

零失敗技巧
Successful Cooking Skills

萬用汁在哪裏有售？
萬用汁可以購自日式超市，根據指示調稀便可以。

Where can I get the concentrated dashi?
Just get it from Japanese supermarket and dilute it according to the instructions.

菠菜要擠乾水分嗎？
菠菜一定要擠乾水分，否則紫菜會變軟。

Should I squeeze dry the spinach?
Yes. You must squeeze dry the spinach. Otherwise, the Nori seaweed will pick up the moisture and turn soggy.

香辣銀魚長通粉

Penne with Spicy Anchovy Fillet

材料（4 人份量）

長通粉 1 杯
銀魚柳 1/2 罐
紅辣椒各 1 隻
番茄乾數隻
黑橄欖數粒
香草意粉醬 3/4 杯
蒜茸 2 茶匙
洋葱絲 1/4 個份量
橄欖油 2 湯匙

調味料

鹽適量
黑椒碎少許

做法

1. 長通粉放入滾水中，加入少許鹽和橄欖油煮約 5 分鐘，盛起。
2. 番茄乾用暖水浸片刻，辣椒去籽，辣椒與番茄乾分別切碎。
3. 銀魚柳略壓碎。
4. 燒熱橄欖油 1 湯匙，放入蒜茸、洋葱絲、辣椒，炒香後加入番茄乾、銀魚柳再炒片刻。
5. 加入長通粉炒勻，下意粉醬、黑橄欖，炒勻，加入調味料，炒勻便可上碟，享用時灑些芝士粉味道會更佳。

Ingredients (Serves 4)

1 cup penne
1/2 can anchovy fillet
1 red chilli
a few dried tomatoes
a few black olives
3/4 cup pasta sauce with herbs
2 tsp finely chopped garlic
1/4 onion, shredded
2 tbsp olive oil

Seasoning

a pinch of salt
a pinch of ground black pepper

Method

1. Put penne in boiling water. Add a pinch of salt and a dash of olive oil. Cook for 5 minutes. Remove.
2. Soak dried tomatoes in warm water for a while. Seed red chilli. Finely chop red chilli and dried tomatoes separately.
3. Crush anchovy fillet.
4. Heat 1 tbsp of olive oil. Put in garlic, onion and red chilli. Stir fry until fragrant. Add dried tomatoes and anchovy fillet. Stir fry for a while.
5. Put in penne and stir fry. Add pasta sauce and black olives. Stir fry. Put in seasoning. Stir fry and dish up. If served with grated parmesan cheese, the penne will taste better.

零失敗技巧
Successful Cooking Skills

可以用其他意大粉代替長通粉嗎？

除了用長通粉，還可以選用螺絲粉、蝴蝶粉等。

Can I use other pasta instead of penne?

Yes. You can use fusilli, rotini or farfalle instead.

甚麼是油浸鯷魚柳？

油浸鯷魚柳味道鹹香，入口幾乎會在嘴裏融化。因為它的味道偏鹹，要試味才下鹽調味。

What are anchovies in oil?

They are small fish which are salted and canned or bottled in oil. They taste salty and flavourful. They are so tender that they melt in your mouth. However, they tend to be salty and you should taste the dish before seasoning any further.

紅辣椒去籽與否的分別大嗎？

如不剔去紅辣椒籽，長通粉的味道會較辛辣。宜用刀剩開紅辣椒，再用刀刮去辣椒籽，勿用手指碰觸辣椒籽，否則手指會灼痛感，久久不會消退。

Is it really necessary to de-seed the chillies?

If you use whole chillies with the seeds, the penne would taste a lot spicier. To de-seed the chillies, cut along the length with a knife and scrape off the seeds with the knife. Try not to touch the seeds with your fingers. Otherwise, the burning sensation would stay for a long time.

椰汁糕

Coconut Pudding

材料（4 人份量）

大菜 2 克
魚膠粉 2 1/2 湯匙
糖 90 克
椰汁 200 克
鮮奶 120 毫升
蛋白 2 個（或用蛋白粉）
水 1 杯

做法

1. 大菜洗淨剪碎，用水 1 杯浸軟後開大火煮溶。
2. 糖與魚膠粉拌勻，加入大菜溶液，攪勻至完全溶解攤凍。
3. 加入鮮奶、椰汁攪勻。
4. 蛋白打起，加入上述混合溶液拌勻（可坐放於冰上），然後倒入容器內，放入雪櫃雪至凝固。

Ingredients (Serves 4)

2 g agar-agar
2 1/2 tbsp gelatine
90 g sugar
200 g coconut milk
120 ml milk
2 egg whites (or equivalent amount of powdered egg white)
1 cup water

Method

1. Cut the agar-agar into short lengths with a pair of scissors. Soak them in 1 cup of water in a pot until soft. Boil over high heat until they dissolve.
2. Mix gelatine and sugar together. Add hot agar-agar solution from step 1. Stir until gelatine and sugar dissolve. Leave it to cool.
3. Add milk and coconut milk. Stir well.
4. Whisk the egg whites until stiff. Add the coconut milk mixture to the egg whites. Fold well (optionally, on top of an ice bath). Pour the resulting mixture into a container. Refrigerate until set.

零失敗技巧
Successful Cooking Skills

怎樣可以避免椰汁糕分開兩層？

將容器坐於冰上才拌入蛋白，可使椰汁糕快些凝固，同時避免輕和容易浮面的蛋白，將凍糕分成兩層。

How do you prevent the coconut pudding from separating?

Make a batter in a metal bowl and put the bowl over an ice water bath before stirring in the egg white. The batter will set more quickly this way so that the egg white which is lighter and tends to float doesn't have time to separate from the rest of the batter.

用生蛋白安全衛生沒有保障，怎麼辦？

可以用蛋白粉它較為安全。若使用蛋白粉，請遵照說明書的做法和份量。

I'm concerned about eating raw egg white which could be a food safety issue. What other options do I have?

You can use powdered egg white instead as it is safer. If you use powdered egg white, follow the instructions on how to rehydrate it. Also calculate the equivalent weight for this recipe.

黑芝麻奶凍

Black Sesame Panna Cotta

材料

鮮奶 1 杯
忌廉 200 毫升
糖 80 克
魚膠粉 1 1/2 湯匙
黑芝麻醬 2 湯匙

Ingredients

1 cup milk
200 ml whipping cream
80 g sugar
1 1/2 tbsp gelatine
2 tbsp black sesame paste

做法

1. 鮮奶、忌廉、黑芝麻醬拌勻。
2. 魚膠粉與糖拌勻，加入上述材料中。
3. 用慢火將以上的材料煮熱後，分別舀進小杯內，放入雪櫃雪至凝固即可享用。

Method

1. Mix milk, whipping cream and black sesame paste together until lump-free.
2. Mix gelatine with sugar. Put them into the black sesame mixture from step 1.
3. Cook the resulting mixture over low heat until it boils. Divide among small containers. Refrigerated until set.

零失敗技巧
Successful Cooking Skills

為甚麼選用黑芝麻醬？

因為它較濃縮、味道較香，兼省工夫。

Why do you use black sesame paste instead of grinding the seeds from scratch?

Black sesame paste is concentrated and is very aromatic. It also saves you much work.

為甚麼只將混合料煮熱而不是煮滾呢？

如將混合料煮滾，會令油脂分離，奶凍入口時較粗糙。

Why is the mixture not boiled, but only heated?

If you boil the mixture, the grease will separate. The panna cotta will be lumpy and not velvety after set.

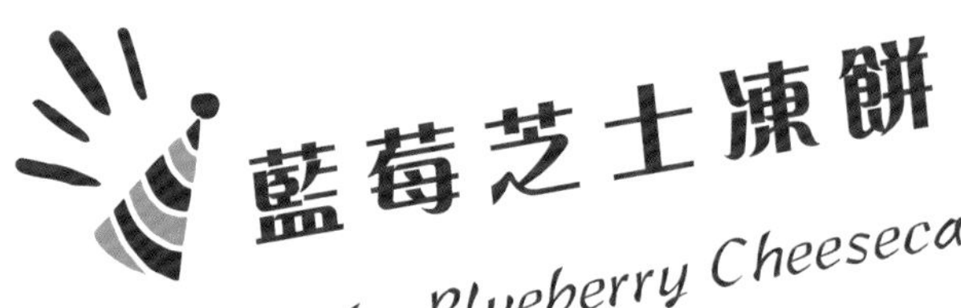

藍莓芝士凍餅
No-bake Blueberry Cheesecakes

材料（8 個份量）
忌廉芝士 180 克
砂糖 60 克
鮮奶 45 毫升
檸檬汁 1 湯匙
魚膠粉 2 茶匙
水 3 湯匙
忌廉半杯
藍莓醬 3 湯匙

餅底
消化餅 100 克
溶化牛油 3 湯匙

餅面
藍莓餡料 4 湯匙

做法
1. 消化餅壓碎，加入溶化牛油拌勻，放入已墊紙的鬆餅模內。
2. 忌廉打發；魚膠粉用水拌勻，坐於熱水至溶，備用。
3. 忌廉芝士打至滑，加入砂糖拂打，再下鮮奶拌打。
4. 加入魚膠粉水及檸檬汁拌勻，最後輕輕拌入忌廉。
5. 將適量芝士材料放於餅底上，鋪上藍莓醬，再傾入芝士材料，冷藏至凝固。
6. 餅面澆上適量藍莓餡料即可。

自製藍莓餡料

材料
藍莓 1 盒
砂糖 2 湯匙
檸檬汁 1 茶匙
粟粉水適量

做法
藍莓用水半杯煮約 3 分鐘，加入糖及檸檬汁煮勻，最後下粟粉水煮至稠。

Ingredients (Makes 8 cakes)
180 g cream cheese
60 g sugar
45 ml milk
1 tbsp lemon juice
2 tsp gelatine powder
3 tbsp water
1/2 cup whipping cream
3 tbsp blueberry jam

Crust
100 g digestive biscuits
3 tbsp melted butter

Topping
4 tbsp blueberry jam

Method

1. To make the crust, crush the digestive biscuits finely. Add melted butter and stir well. Divide evenly among 8 muffin tins lined with parchment paper.
2. Beat the whipping cream until stiff. Set aside. Mix the gelatine in water and heat it up over a pot of simmering water. Stir until gelatine dissolves completely.
3. Beat the cream cheese until smooth. Add sugar and beat well. Add milk and beat again.
4. Add lemon juice and the gelatine solution from step (2). Gently fold in the whipped cream from step (2) at last.
5. Pour some cheese mixture into the muffin tins over the crust. Top with blueberry jam. Pour the remaining cheese mixture over the blueberry jam. Refrigerate until set.
6. Dribble some blueberry jam on top. Serve.

變化

用 6 吋圓形蛋糕模代替鬆餅模

Variations

Use a 6-inch round cake mould instead of muffin tins.

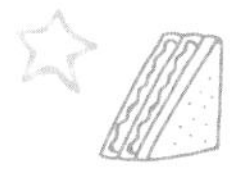

Homemade blueberry jam

Ingredients

1 box fresh blueberries
2 tbsp sugar
1 tsp lemon juice
cornflour solution

Method

Cook the blueberries in 1/2 cup of water for 3 minutes. Add sugar and lemon juice. Cook further and mix well. Thicken with cornflour solution at last and cook until thick.

零失敗技巧
Successful Cooking Skills

有甚麼方法令忌廉容易打發？

預先雪凍打發忌廉的玻璃盆或不鏽鋼容器，或將容器放在冰塊上，這樣就很容易將忌廉打發至所需的狀態。

Is there any way to beat the whipping cream stiff more quickly?

Chill a glass or stainless steel bowl in advance. Or, put the bowl on ice cubes. That would shorten the time required to beat the cream.

可以用其他果醬代替藍莓醬嗎？

可以用草莓、覆盆子果醬代替，但宜購買內有果粒的。

Can I use other jam instead of blueberry jam?

You can use strawberry or raspberry jam instead. But I prefer those with chunks of fruit inside.

洛神花果凍

Hibiscus Tea Jelly

材料

洛神花 5 朵
糖 3 湯匙
寒天粉 2 湯匙
水 2 杯
芒果 1 個
奇異果 1 個
火龍果 1/2 個
草莓數粒

做法

1. 各種生果切粒；洛神花洗淨；寒天粉與糖先拌勻。
2. 煮滾 2 杯水，1 杯注入洛神花內焗 15 分鐘，另一杯與寒天粉拌勻。
3. 將洛神花茶與寒天粉水拌勻，加入雜果粒。
4. 分放舀進小杯內，雪至凝固。

Ingredients

5 dried hibiscus flowers
3 tbsp sugar
2 tbsp Kanten powder
2 cups water
1 mango
1 kiwifruit
1/2 dragon fruit
a few strawberries

Method

1. Dice all fruits. Rinse the hibiscus flowers. Mix the Kanten powder with sugar.
2. Boil 2 cups of water. Pour 1 cup of boiling water into a cup with the hibiscus flowers and cover the lid. Leave them for a while. Mix the other cup of boiling water with the Kanten and sugar mixture. Stir well.
3. Mix together the hibiscus tea and the Kanten solution. Stir in the diced fruits.
4. Scoop the resulting mixture into small serving cups. Refrigerate until set. Serve.

零失敗技巧
Successful Cooking Skills

寒天粉怎樣才較易溶化？？
寒天粉先與糖拌勻才注入滾水，寒天粉會較容易溶化。

How can I dissolve the Kanten powder more quickly?

You can mix the Kanten powder with sugar first before pouring in boiling water. The Kanten powder would dissolve more easily that way.

可以配搭其他生果嗎？
可隨意配搭生果，例如桃、覆盆子等。

Can I add other fruit to this dessert?

Yes, you can add any fruit you want, such as peaches or raspberries.

黑芝麻糖番薯

Candied Sweet Potato with Black Sesames

材料（4 個份量）

日本番薯 300 克
炒香黑芝麻 2 茶匙
糖 90 克
水 1/3 杯

做法

1. 日本番薯洗淨。
2. 原條番薯焓約 10 分鐘至剛脸，盛起待涼。
3. 番薯用滾刀法切小件，放入平底鑊烘片刻至乾身。
4. 用另一個小煲，將糖、水煮到濃稠，倒入番薯，拌匀，上碟，灑黑芝麻。

Ingredients (Serves 4)

300 g Japanese sweet potatoes
2 tsp toasted black sesames
90 g sugar
1/3 cup water

Method

1. Rinse the sweet potatoes.
2. Steam the sweet potatoes in whole for 10 minutes until just tender. Leave them to cool.
3. Cut the sweet potatoes into random wedges by rolling them slightly after each cut. Place them into a dry pan and fry until dry.
4. In another pot, cook sugar and water until thick and syrupy. Put in the sweet potato wedges. Toss well and transfer onto a serving plate. Sprinkle black sesames on top. Serve.

零失敗技巧
Successful Cooking Skills

因為是連皮食，番薯怎樣才清洗乾淨？

基本上，日本番薯在售賣前已洗乾淨泥污，你只須沖去表面灰塵就可以。如仍怕不乾淨，將番薯放在水喉下，再用牙刷輕擦就成。

The sweet potatoes are cooked with their skin on. How can I ensure the skin is thoroughly cleaned?

Basically, Japanese sweet potatoes are cleaned and washed before packaged. You only need to rinse off the dust on the skin before use. If you're still concerned, you may rinse them under a running tap and scrub them gently with a toothbrush.

怎樣才知道糖水是否濃稠？

將糖水煮至起泡，或用匙羹舀起，看看糖水是否黏着匙羹。

How do I know if the syrup is thick enough?

Boil the syrup until it bubbles. Or, you can scoop it up with a spoon and check if it coats the spoon well.

甜薯銅鑼燒

Sweet Potato Layered Dorayaki

蛋糕材料 （可做 6 個）

雞蛋 2 個
糖 50 克
麵粉 50 克
發粉 1/4 茶匙
溶化牛油 2 湯匙

餡料

番薯（去皮計）100 克
糖 1 湯匙
煉奶、牛油各 1/2 湯匙

餡料做法

1. 番薯去皮切薄片，隔水蒸腍。
2. 將餘下的材料加入，拌勻，便成甜薯餡。
3. 將甜薯餡用保鮮紙包好壓薄，切成約 8 厘米 x 8 厘米的方塊，待用。

蛋糕做法

1. 麵粉與發粉同篩勻。
2. 蛋與糖打透至呈奶白色，篩入麵粉，輕輕拌勻，注入溶化牛油，拌勻。
3. 平底鍋內抹油，先煎兩片約 10 厘米的圓形餅皮，在其中一片放上甜薯餡（圖 1），再放上餅皮（圖 2）。
4. 再將麵糊舀進平底鍋內（大小如做法 3），放上甜薯餡，將做法（3）夾餅放上（圖 3），再烘片刻，然後用手拿着夾餅垂直放於鍋邊，烘至餅邊全熟便可以享用（圖 4）。
5. 其餘粉漿和甜薯餡均依做法（3、4）完成，約可製成 6 個甜薯銅鑼燒。

Pancake batter (makes 6 Dorayaki)

2 eggs
50 g sugar
50 g flour
1/4 tsp baking powder
2 tbsp melted butter

Filling

100 g sweet potatoes (peeled)
1 tbsp sugar
1/2 tbsp condensed milk
1/2 tbsp butter

Method

Filling

1. Slice the sweet potatoes thinly. Steamed until soft.
2. Put in the remaining filling ingredients and mix well. This is the filling.
3. Put filling on a cling wrap. Cover with another sheet of cling wrap and press evenly. Cut into 8-cm squares. Set aside.

1

2

3

4

Pancake

1. Sieve the flour and baking powder together.
2. Beat eggs with sugar until pale. Sieve in the flour and baking powder. Fold gently to mix well. Stir in melted butter. Mix well.
3. Grease and heat up a pan. Pour in 2 small ladles of batter with some space between them. Make 2 pancakes about 10 cm in diameter. Put a slice of sweet potato filling on one of them (fig.1). Flip the other pancake over to cover the filling (fig.2). Set aside.
4. Pour a small ladle of batter into the pan. The pancake should be about 10 cm in diameter. Place a slice of sweet potato filling over the pancake before it sets. Then put the pancake sandwich from step 3 over the half-set pancake (fig.3). Fry briefly until set. Then hold the sandwich with your hand to brown it lightly on all sides (fig.4).
5. Repeat method 3 and 4 with the remaining batter and filling. The recipe makes 6 layered Dorayaki.

零失敗技巧 Successful Cooking Skills

要用甚麼火候烘煎餅？

炮製這夾餅要有耐性和使用弱火，慢慢的一層一層烘煎，必能成功。

When I fry the pancakes, what heat should I use?

It takes patience to fry the pancakes. Just cook it over low heat and do it one by one. You're more likely to pull it off that way.

陳皮紅豆沙

Red Bean Soup with Dried Tangerine Peel

材料（4 人份量）

紅豆 300 克
陳皮 2 角
片糖 1-2 片

做法

1. 紅豆洗淨。煮滾 2 杯水，加入紅豆煮約 10 分鐘，關火焗片刻約 20 分鐘，倒去水。
2. 陳皮用水浸軟，刮去果瓤。
3. 燒滾 6 杯水，加入紅豆、陳皮，先用大火煮約 10 分鐘，再轉中火煮約 1 小時至起沙。
4. 加入片糖，待片糖煮溶便可以享用。

Ingredients (Serves 4)

300 g red beans
2 quarters dried tangerine peel
1 to 2 raw cane sugar slabs

Method

1. Rinse the red beans. Boil 2 cups of water in a pot. Put in the red beans and cook for 10 minutes. Turn off the heat and cover the lid. Leave it for 20 minutes. Drain.
2. Soak the dried tangerine peel in water until soft. Scrape off the pith.
3. Boil 6 cups of water. Put in the red beans and dried tangerine peel. Boil over high heat for 10 minutes first. Turn to medium heat and cook for about 1 hour until the beans are split open and the sweet soup turns starchy.
4. Add sugar slabs. Cook until they dissolve. Serve.

零失敗技巧
Successful Cooking Skills

紅豆較難煲腍，有解決方法嗎？

先將紅豆浸半天（期間要換水），會較易煲軟紅豆。

It takes forever to cook the red beans. Is there any trick to speed up the process?

Soak the red beans in water for 12 hours before use. Drain and replenish with freshwater once or twice. The red beans will turn tender more quickly after being saturated with water.

陳皮是否年份越久越香？

是的，但年份久的陳皮較昂貴，選用十年陳皮香味已足夠。

People say the older the dried tangerine peel, the more fragrant it is. Is it true?

Yes, it's true. However, aged dried tangerine peel could be very expensive. For this recipe, pick those about 10 years old would be good enough.

提子麵包布甸

Raisin Bread Pudding

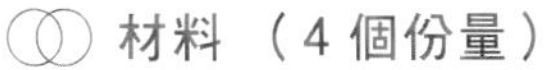

材料（4 個份量）

提子方包 6 片
牛油 30 克
雞蛋 1 個
鮮奶 2/3 杯
砂糖 2 湯匙
橙酒 1/2 湯匙
橙皮茸 2 茶匙
豆蔻粉 1/4 茶匙
提子乾、金提子乾共 1/4 杯

塗面料（坐溶）

黃梅果醬 1 湯匙
水 1 湯匙

做法

1. 提子方包切去硬皮，塗上牛油後烘至少許金黃色，切小塊。
2. 雞蛋及鮮奶拌勻，加入砂糖、橙酒、橙皮茸及豆蔻粉拌勻。
3. 鬆餅盆內塗油或墊上牛油紙，放上一層提子包，灑上提子乾，再鋪上餘下的麵包，注入步驟（2）的材料。
4. 放入預熱焗爐，以 190℃焗約 20 分鐘至金黃色，待涼。
5. 最後掃上黃梅果醬品嘗。

Ingredients (Makes 4 puddings)

6 slices raisin sandwich bread
30 g butter
1 egg
2/3 cup milk
2 tbsp sugar
1/2 tbsp Cointreau
2 tsp grate orange zest
1/4 tsp ground cardamom
1/4 cup dried raisins and golden sultana

Glazing: (mixed together and melted over a pot of simmering water)

1 tbsp apricot jam
1 tbsp water

Method

1. Cut off the crust of the raisin sandwich bread. Spread butter on them and bake in an oven until golden. Cut them into small pieces.
2. Whisk milk and eggs together. Add sugar, Cointreau, grated orange zest and ground cardamom. Stir well.
3. Grease the muffin tins or line them with parchment paper. Put a layer of raisin sandwich bread on the bottom and top with some dried raisins and golden sultana. Arrange the remaining bread over the raisins and sultana. Pour the egg mixture from step (2) over the bread slowly.
4. Bake in a preheated oven at 190°C for 20 minutes until golden. Leave them to cool.
5. Brush the apricot glazing on top at last. Cut into pieces and serve.

變化

將布甸料舀進長方形餅模（8 吋 x3 吋 ）內，以 190℃焗約 25 分鐘即可。

Variations

Use a 8" x 3" rectangular baking tray instead. Bake at 190˚C for 25 minutes.

零失敗技巧
Successful Cooking Skills

食材可以用甚麼變化？

這款甜品變化無窮，例如提子包可以其他甜麵包代替；提子乾換上紅莓乾等。

Is there any variation to this recipe?

There are endless variations to it. You can use other sweet bread instead of the raisin loaf. You can use dried cranberries instead of dried raisins. Feel free to improvise.

有甚麼美味竅門？

塗上牛油的麵包烘焙時香氣四溢，口感酥脆；切記不要用人造牛油。

Is there any trick to a tasty bread pudding?

One trick is to always use butter to spread on the bread. Do not use margarine. Only butter can give it the crispy texture and divine aroma when baked.

楓漿雪芳蛋糕

Maple Chiffon Cake

材料（長方形餅模1個（8吋x3吋）

麵粉 100 克
發粉 1 1/4 茶匙
鹽 1/4 茶匙
砂糖 100 克
菜油 45 毫升
蛋黃 2 個
鮮奶 45 毫升
楓漿 3 湯匙
雲呢拿香油 1/3 茶匙
蛋白 100 毫升
他他粉 1/4 茶匙

做法

1. 麵粉、發粉及鹽同篩勻，加入糖拌勻。
2. 菜油、蛋黃、鮮奶、楓漿及雲呢拿香油拌勻，傾入做法（1）的材料拌勻（圖 1-2）。
3. 蛋白打至企身，加入他他粉拌勻，再傾入做法（2）的材料中拌勻（圖 3-4）。
4. 餅盆塗油或墊上牛油紙，傾入混合物（圖 5），放入預熱焗爐用 190℃焗約 35 分鐘即可。

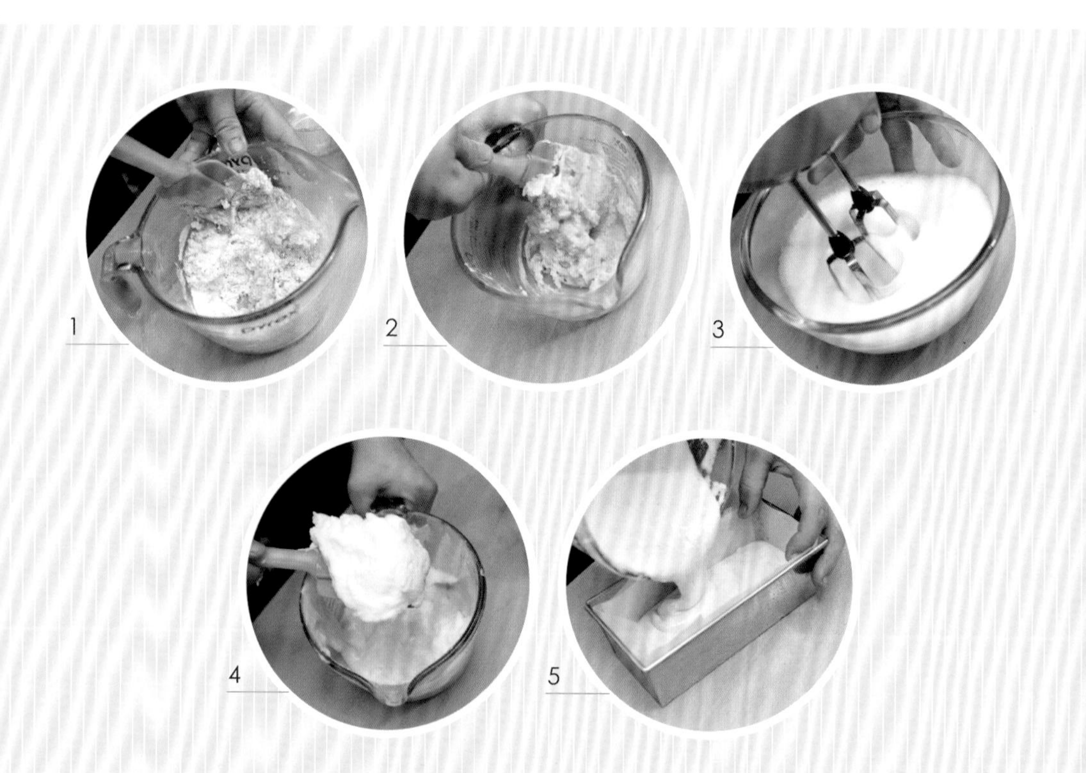

Ingredients (Makes one 8" x 3" rectangular cake)

100 g flour
1 1/4 tsp baking powder
1/4 tsp salt
100 g sugar
45 ml vegetable oil
2 egg yolks
45 ml milk
3 tbsp maple syrup
1/3 tsp vanilla essence
100 ml egg whites
1/4 tsp cream of tartar

Method

1. Sieve flour, baking powder and salt together. Add sugar and mix well.
2. In a separate mixing bowl, mix vegetable oil, egg yolks, milk, maple syrup and vanilla essence. Then stir in the dry ingredients from method (1) and mix well (fig.1-2).
3. In another bowl, beat egg whites until stiff. Add cream of tartar and stir well. Then fold it into the flour mixture from method (2) (fig.3-4).
4. Grease or line a loaf tin with parchment paper. Pour the batter in (fig.5) and bake in a preheated oven at 190°C for 35 minutes.

零失敗技巧
Successful Cooking Skills

為甚麼要先篩勻麵粉、發粉及鹽？

除了可去除雜質、結塊外，也可以令發粉和鹽分佈均勻，對蛋糕是否膨脹均勻和味道是否一致有很大影響。

Why do you sieve the flour with the baking powder and salt together at first?

That would remove any lump or impurities in the dry ingredients. It also helps distribute the salt and baking powder evenly in the flour. This is very important to the consistency of the cake both in terms of flavour and airiness.

蛋白與其他材料拌勻時，有甚麼技巧？

蛋白打至企身，與其他材料拌勻時，宜用橡皮刮刀從底部向上微拌。

When I mix the meringue with other ingredients, is there anything that needs my attention?

Beat the egg whites till stiff. Then add the meringue to the other ingredients and fold the meringue in by scrapping the rubber spatula along the bottom of the bowl and lifting the mixture gently until well incorporated.

番茄西瓜汁

Tomato Watermelon Juice

材料

番茄 3 個
無核西瓜 1/4 個

做法

1. 洗淨番茄，去蒂，切碎。
2. 西瓜起肉，切碎。
3. 將材料放入榨汁機內，榨出汁液，即可享用。

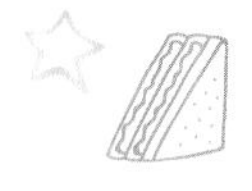

Ingredients

3 tomatoes

1/4 seedless watermelon

Method

1. Rinse the tomatoes and remove the stems. Cut into pieces.
2. Peel the watermelon. Cut into pieces.
3. Press all ingredients through an electric juicer. Save the juice and serve.

零失敗技巧
Successful Cooking Skills

份量有多少？

大約有 400 毫升，可以按聚會人數增加食材份量。這果汁在西瓜味中有絲絲番茄香，配搭新穎。

How much juice does this recipe make?

It makes 400 ml of juice blend roughly. Just multiply the amount of ingredients to cater to bigger parties. You can taste the tartness of tomato faintly in the watermelon juice. It tastes interesting and yummy.

可以同攪拌機代替榨汁機嗎？

可以的，但建議先去除番茄籽和皮，以免飲時滿口渣滓。

Can I use a blender instead of a juicer?

Yes, you can. But you should de-seed and skin the tomatoes first. Otherwise, the juice blend won't be smooth and your guests have to spit out the seeds and skin.

檸檬薄荷茶

Mint Lemon Tea

材料

檸檬 1 個
薄荷葉 20 片
蜂蜜 2 湯匙
沸水 3 杯
冰粒少許
薄荷葉、檸檬各數片（作裝飾）

做法

1. 檸檬榨汁。
2. 洗淨薄荷葉，撕碎，加入沸水焗 5 分鐘，隔渣。
3. 加入蜂蜜及檸檬汁，拌勻。
4. 飲時加入薄荷葉、檸檬片和冰粒。

Ingredients

1 lemon
20 mint leaves
2 tbsp honey
3 cups boiling hot water
ice cubes
mint leaves and lemon slices (as garnish)

Method

1. Squeeze the lemon.
2. Rinse the mint leaves and tear them into small pieces. Add boiling water and cover the lid. Leave them for 5 minutes. Strain.
3. Add honey and lemon juice to the mint tea. Stir well.
4. Add mint leaves, a slice of lemon and ice cubes before serving.

零失敗技巧
Successful Cooking Skills

份量有多少？

大約有 750 毫升，可以按聚會人數增加食材份量。喝時滿口薄荷清新，除煩去悶的良方。

How much tea does this recipe make?

It makes roughly 750 ml. You can size it up by multiplying the amount of ingredients used for bigger parties. It tastes refreshing and is a good way to invigorate the spirits.

為甚麼要先撕碎薄荷葉？

經撕碎的薄荷葉，會更加出味。

Why do you tear the mint leaves finely first?

That helps release the essential oil in the mint leaves, so that the tea would carry stronger mint flavour.

百香果蜜汁

Honey Passionfruit Drink

材料
百香果 6 個
蜂蜜 2 湯匙
暖水 1 1/2 杯

做法
1. 暖水和蜂蜜調勻。
2. 百香果切半，舀起果汁和籽。
3. 將蜂蜜水、百香果汁和籽調勻，便可飲用。

Ingredients
6 passionfruits
2 tbsp honey
1 1/2 cups warm water

Method
1. Stir honey into the warm water.
2. Cut the passionfruits into halves. Scoop out the pulp, seeds and the juice.
3. Put the passionfruit pulp, seeds and juice into the honey-water mixture. Stir well and serve.

零失敗技巧
Successful Cooking Skills

份量有多少？
大約有 500 毫升，可以按聚會人數增加食材份量。

How much juice does this recipe make?
It makes 500 ml of juice roughly. Just multiply the amount of ingredients to cater to bigger parties.

不太喜歡百香果籽，怎樣除去它？
用篩子隔去就可以，隔時宜用匙略壓百香果籽。

I don't like the seeds of passionfruit. How can I remove them?
Just pass the pulp through a wire mesh to remove it. Then gently squeeze the pulp with a spoon to extract the juice.

適宜作餐前飲品嗎？
它有濃濃的百香果香，味道討好，對皮膚有滋潤作用，最適合女性聚會時飲用。

Is it a good pre-meal drink?
Yes, it is. The fruity and tangy drink is guaranteed to be an instant hit in all-girl parties. It also helps keep your skin supple, ladies.

三莓汁

Triple Berry Blend

材料

覆盆子、藍莓各 1/2 盒
蔓越橘汁 1/2 杯
純味乳酪、粟米片各 1 杯
蜂蜜 1 湯匙
冰塊少許

做法

1. 洗淨覆盆子、藍莓。
2. 將所有材料和 1/2 杯粟米片放入攪拌機內打至幼滑。
3. 倒入杯內，放入餘下的粟米片即可享用。

Ingredients

1/2 box raspberries
1/2 box blueberries
1/2 cup cranberry juice
1 cup plain yoghurt
1 cup cornflakes
1 tbsp honey
ice cubes

Method

1. Rinse the raspberries and blueberries.
2. Put all ingredients and 1/2 cup of cornflakes into a blender. Blend until smooth.
3. Pour into a serving glass and put the remaining cornflakes on top and serve.

零失敗技巧
Successful Cooking Skills

份量有多少？

大約有 350 毫升，是一杯份量，可以按聚會人數增加食材份量。這果汁充滿清新的覆盆子香氣，面灑上脆脆的粟米片，是 Party 飲品清新之選。

How much juice does this recipe make?

It makes about 350 ml, which is about the size of a highball glass. You can multiply the amount of ingredients for bigger parties. This juice is full of berry goodies dominated by tart raspberry taste. Sprinkle crunchy cornflakes on top for a refreshing party drink.

這果汁可以預先攪拌嗎？

最好是攪拌後隨即飲用，否則營養容易流失，同時奶類製品容易變壞。

Can I make this juice ahead of time?

No. It works best if you can serve it right after blending it. First off, the nutritional value may be compromised if let stand for too long. Secondly, dairy products in the juice may also go stale if made in advance.

烏豆糙米茶

Brown Rice Tea with Black Beans

材料

烏豆 60 克
糙米 60 克
生薏米 60 克
熟薏米 30 克
水 6 杯

做法

1. 洗淨各材料。
2. 燒滾水，放入所有材料，滾後轉中火再煲約 1 小時便可以飲用。

Ingredients

60 g black beans
60 g brown rice
60 g raw job's tears
30 g fried job's tears
6 cups water

Method

1. Rinse all ingredients.
2. Boil water. Put in all ingredients. Bring to the boil and turn to medium heat. Cook for 1 hour and serve.

零失敗技巧
Successful Cooking Skills

份量有多少？

大約有 1 公升。這茶飲有米香味，入口清新，不膩滯，是 Party 的有益茶水。

How much tea does this recipe make?

It makes about 1 litre of tea. This tea carries a hint of rice fragrance with a refreshing palate. It's a healthful party drink that also aids indigestion.

鮮橙香草茶

Orange Herbal Tea

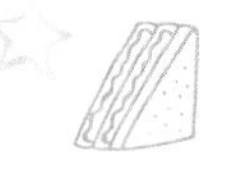

材料

百里香 2 棵
鼠尾草葉 6 片
沸水 1 杯
蜂蜜 1 湯匙
橙 1 個

做法

1. 洗淨百里香、鼠尾草葉，切碎，加入沸水焗 5 分鐘。
2. 拌入蜂蜜，再焗 5 分鐘，隔渣。
3. 橙切半，榨汁，與香草蜂蜜茶調勻便可享用。

Ingredients

2 sprigs thyme
6 sage leaves
1 cup boiling hot water
1 tbsp honey
1 orange

Method

1. Rinse the thyme and sage leaves. Finely chop them. Transfer to a cup and soak them in boiling hot water with the lid covered for 5 minutes.
2. Stir in honey. Cover the lid again for 5 minutes. Strain.
3. Cut the orange in half. Squeeze it and pour the juice into the honey herbal tea from method 2. Serve.

零失敗技巧
Successful Cooking Skills

份量有多少？
大約有 250 毫升。在聚會的尾聲時泡給客人飲用，留下甜美的回憶。

How much tea does this recipe make?
It makes about 250 ml. You can serve this at the end of the party to leave a sweet lasting impression on your guests.

可以用乾香草嗎？
宜用新鮮的。新鮮的香草味道芳香，乾貨會有點木渣味。

Can I use dried herbs instead?
For this recipe, I recommend using fresh herbs because of their fragrance. Dried herbs carry a woody taste that doesn't work well here.

編者
Forms Kitchen編輯委員會

Editor
Editorial Committee, Forms Kitchen

美術設計
馮景蕊

Design
Carol Fung

美術設計
何秋雲

Typography
Sonia Ho

出版者
香港鰂魚涌英皇道1065號
東達中心1305室
電話
傳真
電郵
網址

Publisher
Forms Kitchen
Room 1305, Eastern Centre, 1065 King's Road,
Quarry Bay, Hong Kong.
Tel: 2564 7511
Fax: 2565 5539
Email: info@wanlibk.com
Web Site: http://www.wanlibk.com
http://www.facebook.com/wanlibk

發行者
香港聯合書刊物流有限公司
香港新界大埔汀麗路36號
中華商務印刷大廈3字樓
電話
傳真
電郵

Distributor
SUP Publishing Logistics (HK) Ltd.
3/F., C&C Building, 36 Ting Lai Road,
Tai Po, N.T., Hong Kong
Tel: 2150 2100
Fax: 2407 3062
Email: info@suplogistics.com.hk

承印者
中華商務彩色印刷有限公司

Printer
C & C Offset Printing Co., Ltd.

出版日期
二零一九年三月第一次印刷

Publishing Date
First print in March 2019

Published in Hong Kong by Forms Kitchen,
a division of Wan Li Book Company Limited.
ISBN 978-962-14-6989-2

鳴謝以下作者提供食譜（排名不分先後）：
黃美鳳、Feliz Chan、Winnie姐